BUSES, BANKERS
&
THE BEER OF REVENGE

BUSES, BANKERS
&
THE BEER OF
REVENGE

An Eccentric Engineer Collection

Justin Pollard

The Institution of Engineering and Technology

Published by The Institution of Engineering and Technology, London, United Kingdom

The Institution of Engineering and Technology is registered as a Charity in England & Wales (no. 211014) and Scotland (no. SC038698).

First published 2012

The Institution of Engineering and Technology
Michael Faraday House
Six Hills Way, Stevenage
Herts, SG1 2AY, United Kingdom

www.theiet.org

British Library Cataloguing in Publication Data
A catalogue record for this product is available from the British Library

ISBN 978-1-84919-581-2 (hardback)
ISBN 978-1-84919-582-9 (electronic)

Typeset in the UK by Carnegie Book Production, Lancaster
Printed in the UK by MPG Books Group

For

Martyn, Helen, Emma, Ceri and Euan

aut viam inveniam aut faciam

Contents

Foreword

We are very pleased to present to readers this collection of Justin Pollard's 'Eccentric Engineer' columns published in *E&T* magazine.

Justin is one of Britain's top historians of science and technology, assistant producer of BBC TV's award-winning *QI* show, author of many books and history consultant for a number of blockbuster Hollywood movies. We are honoured to have him among our regular contributors.

'Eccentric Engineer' is always witty, impeccably written and factually impressive. It brings back the little known and totally forgotten personalities and inventions from the past and does it in the form of a gripping and often very funny narrative. Combined with a popular caption competition for the readers, 'Eccentric Engineer' is one of *E&T* magazine's undisputed highlights.

We hope you will enjoy reading (or re-reading) these columns as much as we enjoy editing them.

Engineering eccentrics of all countries, we salute you!

Dickon Ross, Editor-in-Chief

Introduction

Nobody would ever deny that engineering is very important. Everybody wants engineers involved when it comes to building bridges or installing computer networks, but that does not mean the subject has always had a reputation as being wholly exciting either. Perhaps it's because almost everything in our lives is or should be in some way engineered that it, and its practitioners, have something of a reputation for seriousness. Some engineers even whisper amongst friends that people think their job is, well, boring.

That's where the Eccentric Engineer comes in. Since the autumn of 2007 the Eccentric Engineer column has been campaigning to remind engineers of the extraordinary role that their subject, in all its many guises, has played in human history, from the death throes of Archimedes to the origin of Christmas crackers.

This book gathers together three years of wandering through history highlighting not simply the most famous engineering stories but the unusual, the erratic, and occasionally the apparently insane. In its fifty stories taken from the monthly columns in *E&T* magazine, it covers everything from aircraft carriers made of ice, to the origins of the omnibus. We'll toy with Roman turbines, and Greek computers, look at Renaissance hypertext and have arguments with America over the shape of our lightning conductors. We'll shake Scotland with earthquakes and build cars out of beans.

I've arranged the stories into themed chapters although they were never written with a particular order in mind and they can be read in any order you please. They don't form an overall thesis, they are just a series of stories on the sorts of things that tend to get left out of Histories of Engineering and Science and which I for one think are poorer for the lack of them. But it is what lies behind these stories, behind the retort stands and the test benches, behind the trial products and the prototypes that really matters – the inspirational people who for centuries have wrestled ideas from the realms of fantasy into the real world, often whilst being cheerily ridiculed by all those around them.

From ancient Japanese tombs to the Beer of Revenge, the Eccentric Engineer tries to show how these people have created an engineered world, where engineers are not simply the diligent maintenance crews we often take them for, but are at the heart of our most creative enterprise – building the future. Some may use steel, spanners and semi-conductors as their tools, but that doesn't make their job any the less inspirational. Engineering is an art form and this book is a gallery dedicated to their achievement.

I

PROLOGUE

Episode One – A Cautionary Tale – In which we meet a man who disliked being an engineer so much that it killed him

Not everyone wants to be an engineer. Archimedes was one of the greatest engineers in the classical world but would not have thanked you for calling him that. Indeed he managed to embody the very best and worst attitudes towards the profession in a single man. He believed only pure mathematics really mattered and, thanks to his invention of the milometer and an unfortunate incident with a Roman, he paid for that belief with his life. This only goes to show that whilst pure theory might be a higher calling, it's a very poor form of self-defence. Vitruvius, the Roman architect and engineer, tells us that Archimedes had such a low view of engineering that:

... he yet would not deign to leave behind him any commentary or writing on such subjects; but, repudiating as sordid and ignoble the whole trade of engineering, and every sort of art that lends itself to mere use and profit.

This was, in all honesty, a mistake, and his negative attitude would lead to his death – and here's how. Engineering is rather useful, as the Roman army who were then marching towards his home town of Syracuse knew. In fact the man who had really persuaded them of this was Archimedes himself who had invented for them one of the most useful devices in the ancient world and something which is still a fixture in every car and lorry today – the odometer, an example of which is the milometer on every dashboard.

Archimedes' odometer was based on a wooden cart with a series of toothed wheels connecting the axle to a hopper of pebbles, one of which dropped into a collecting bowl with each full turn of the gears. As the circumference of the wheels was known so the distance travelled during one full cycle of the gears could be measured. By multiplying this distance by the number of pebbles in the bowl at the end of the journey the total length of the journey could be calculated. Easy – in fact so easy that Archimedes didn't even think it was worth writing down.

But the Romans did. For them the odometer wasn't a vulgar toy but a device for laying out Roman milestones which meant they knew how far one place was from the other, which in turn meant they knew how long it would take to get an army from one place to another. This enabled them to plan military campaigns over wide areas with great precision – just like the one they were planning right now against Archimedes' home town of Syracuse.

What happened when they arrived shows once again some of the very best and very worst in attitudes towards technology and engineering. Plutarch tells us that Archimedes turned his mind to creating wonderful siege engines which employed the mathematics he so loved to perform apparently superhuman acts. Great catapults were designed to rain down stones upon the legions who foolishly believed themselves to be out of range. When Roman ships approached the harbour walls, cranes were swung out which either dropped huge rocks on their ships to sink them or hooked them out of the water, dumping their crews in the sea before overturning them and consigning them to the depths. Plutarch provides a vivid description of the result on the Roman soldiers' psyches:

... such terror had seized upon the Romans, that, if they did but see a little rope or a piece of wood from the wall, instantly crying out that there it was again, Archimedes was about to let fly some engine at them, they turned their backs and fled ...
Plutarch *Parallel Lives – Life of Marcellus* Chapter 17

In the end however, the relentless might of the Roman army triumphed and as the Roman army flooded into the town, their general Marcellus sent out orders to fetch the world's most eminent engineer, Archimedes, to him. A soldier, so Plutarch tells us, came upon the great man working out a mathematical problem in the sand and ordered him to report to the general. But Archimedes was deep in thought and told him to go away until he had finished – after all he was pondering the workings of the universe and all a Roman would want to talk about was engineering. Now a soldier who has spent the last few weeks having Archimedes' machines rain death down upon him and his friends might be forgiven for being a little short tempered. And indeed he was so short tempered that he simply ran Archimedes through on the spot. Problem solved.

And so Archimedes' own poor opinion of his engineering led to his death. So next time you hear someone being rude about engineering or technology, remind them where such a bad attitude can lead.

After that salutary lesson in respecting your own discipline, let's move on to the main event ...

In which we look at the extraordinary origins of some extraordinary machines and the somewhat eccentric engineers behind them.

Episode Two – A Small Miracle

Just suppose for a moment that you find yourself in ancient Athens in, say, the mid-first century BC. You're minding your own business, strolling through the Roman agora when someone comes up to you and says 'What's the time?' You smile, look down at your watch and then remember that it's the first century BC and watches haven't been invented. So, being a quick-thinking type, you look up to the sky to estimate the position of the sun. But it's dark. You think about using the stars. But it's cloudy. So what exactly was an ancient Athenian to do? Well fortunately he only had to step inside a small octagonal building nearby to find himself looking at the most accurate clock to exist before the thirteenth century AD.

How it got there is thanks in the main to the son of a barber in Alexandria. Until his big idea, telling the time was something you did using a sundial, or perhaps a moondial if you were feeling ambitious. Using these the day was divided into twelve parts but, as the length of daylight changed with the season, this gave rise to the peculiar situation where hours in the winter were much shorter than hours in the summer. Fortunately as time was not yet money, no one was too bothered. But not being able to tell the time at all when there was no moon to cast a shadow at night or the sky was overcast was a real problem and that's where a man with the largely unpronounceable name of Ctesibius came in.

In around 270 BC, Ctesibius was working in his father's barber shop when he had a brilliant idea. Working with hair actually seemed to suit Ctesibius and he'd had several good ideas whilst working there, including inventing the first angle-poise mirror to enable clients to look at the back of their heads. But he'd also been thinking about time. At that date there were timers but no clocks and the most useful sort of timer was the one used in law courts – the *clepsydra* or 'water thief'. This consisted of a hollow metal sphere with holes drilled in it and with a small tube soldered onto it. The sphere could be filled with water through the tube and the water would of course drain away in a certain amount of time through the small holes. This was the amount of time each side had to plead their case – fair and square. The tube only came into play again if there was an interruption. By putting a bung in the end of it the judge could prevent atmospheric pressure pushing down on the surface of the water in the sphere and thus prevent the water from coming out. When proceedings were ready to start again he took the bung out and the water started flowing again. You can try this with an old fizzy drinks bottle with some holes in the bottom. But I wouldn't recommend doing it in court.

The *clepsydra* was a splendid if rather wet timer and soon everyone was complaining that they had no time at all, leading Plato to note that:

lawyers are driven by the clepsydra – *never at leisure ...*

The *clepsydra* is just a timer not a clock but it gave Ctesibius an idea. What if the *clepsydra* was always full? Then the water would flow out at a constant speed as there would be constant head pressure. So he added another water tank above the *clepsydra*. This poured water into the top faster than it could flow out at the bottom. That meant the *clepsydra* was always full and any excess water just overflowed. Then he put another tank below the outflow to catch the water coming out of the *clepsydra*. This tank now filled up at a constant speed, so if a scale was put on the tank and a pointer floated in it then he could measure time constantly. His designs proved so accurate that no better mechanical timepiece would be invented for another 1,800 years.

It was just such a water clock that an enquiring Athenian could find in that diminutive and little known building which still stands in the Roman Agora in Athens and is now known as the Tower of the Winds. The Tower had been designed in the first century BC by Andronicus of

Cyrrhus, who aligned the building precisely north which was a feat in itself if one considers that the magnetic compass wasn't known in Europe until the thirteenth century AD. He then placed a bronze weather vane, in the rather fetching shape of Triton holding a pointer, on the roof which would spin to point to one of eight carvings on the eight sides of the tower, each showing one of the winds with the type of weather it brought – cold from the north, hot and drying from the south and so forth. Each of these faces was also fitted with a gnomon and below the sculpture of the winged wind the hours were marked out in carved lines. But like Ctesibius before him, Andronicus of Cyrrhus was aware that even in Athens the skies were not always clear so he installed a water clock fed by a small tower on the south wall which acted as a cistern and which was, in turn, fed from a water supply up on the Acropolis which towered over the agora. This ran day and night, ensuring that Athenians could tell the time regardless of the hour or the weather.

And so the Greeks became the first people to have their lives run by the ticking, or rather the dripping, of a clock. Not that it made them any happier than it does us. As far back as the third century BC the Roman playwright Plautus was bewailing having his life ruled by shadows and water-thieves, preferring to measure time using a far older system and one which has also thankfully survived to this day:

May the gods destroy that man who first discovered hours and who first set up a sundial here; ... For when I was a boy, my only sundial was my stomach, by far the best and truest of clocks ...

Episode Three – Drawing Water

Nearly every spring we're told by a fresh-faced meteorologist to expect a hot, dry summer. But then we've been told that before. However, on the off-chance that this time the skies remain blue and the sun beats down, I have to give some thought to what to do about my vegetables. Being on a water meter I'm loathe to spend the summer watering them with a hose but then I couldn't bear to see them shrivel and die. So I got to thinking about a bore hole and hence mediaeval robot musicians – not something that will come as any surprise to you I'm sure.

That I should connect the two is all thanks to one remarkable engineer who deserves to be much better known. Al-Shaykh Ra'is al-A'mal Badi' al-Zaman Abu al-'Izz ibn Isma'il ibn al-Razzaz al-Jazari (*c.* 1136–1206), thankfully known as just Al-Jazari, is a rather elusive figure.

We know that he came from the northern part of Mesopotamia and that he was the senior engineer to Nasir Al-Din, the Artuqid king of Diyar Bakr whose dynasty then ruled a large chunk of Northern Iraq, Anatolia and Northern Syria. But what makes Al-Jazari really stand out is that he wrote a book, the intriguing sounding *Book of Knowledge of Ingenious Mechanical Devices* (1206).

Indeed 'ingenious' would be something of an understatement. Al-Jazari's book puts even the likes of Agostino Ramelli to shame, containing a bewildering array of machines to make his masters' lives more convenient and entertaining – fifty devices in all including everything from pumps and fountains to blood-letting devices.

A great builder of automata, he was one of the first engineers to turn his mind to what we today would call the 'domestic appliance'. In particular he was fond of building hand-washing machines which are, of course, very important in Islam. One, the Peacock Fountain, released water through a model peacock's beak which then tripped a series of floats, the first of which made a figure appear from behind a door proffering a bar of soap, and the next another figure with a towel. They just don't make domestic appliances like that any more – I've looked.

Clocks were also a passion. His most sophisticated timepiece, known as the 'Castle Clock', was 3.5 metres tall and, as well as telling the time, also displayed the movement of the constellations of the Zodiac, and the orbits of the sun and moon. On each hour a figure appeared from behind a door, accompanied by a tune from five automaton musicians, making it something akin to the first cuckoo clock. More impressive still, its speed could be altered to track the changes in length of day over the year.

Long summer evenings also gave Al-Jazari's masters a chance to enjoy one of his more eccentric inventions – the robotic floating band. This consisted of a boat holding four automata in the shape of musicians, rowed by ranks of mechanical oarsmen. The machine would row around for half an hour or so on a large lake and then the band would strike up with flutes and drums and cymbals. After a short number they'd stop again and the oarsmen would get back to work. The whole system was operated by a turning drum beneath the deck into which pegs were placed to control the various movements. As the pegs could be repositioned to play different tunes, this made it a very early programmable device.

But what does this have to do with my potatoes? Sadly I don't own a lake or a boatful of automatic musicians but it is thanks to Al-Jazari that I can pump water out of the ground. For all the fiendish complexity of his automata, it is not these that make him great. Hidden away in his discussions of amazing machines lie the first written descriptions of a host of mechanisms we couldn't live without today. He provides the first description of casting metal parts in closed sandboxes, a method for making watertight valves, the static balancing of large pulley wheels and the use of segmental gears – even the lamination of wood to prevent warping.

But it is in his water-raising devices that he makes his greatest contribution. His two-cylinder reciprocating piston suction pump was, and frankly remains, one of the most useful machines for anyone who needs to get water anywhere: from a thirsty vegetable patch to a thirsty city. But it is also much more than this. This was not only a very serviceable pump, but in Al-Jazari's description are the first references to some of the most important ideas in all engineering. In the first place this pump provides the earliest account of a crankshaft, that most essential of devices for turning rotary motion into reciprocal motion (or vice-versa). The modern world just wouldn't have got very far without those. It also contains the first connecting-rod mechanism, another rather vital element to many engines. If that seems impressive let's not forget that it also is the first demonstration of the double-acting principle and the first use of a suction pipe, using atmospheric pressure to force water into a pump by creating a partial vacuum.

In fact a very large number of modern machines simply wouldn't work without the principles first set out by Al-Jazari; indeed, much of the world we take for granted would grind to a halt. And I don't like to think what might happen to my potatoes.

Episode Four – The Key to Success

We all know you need to advertise to be successful, or at the very least it helps. But do engineers, be they electrical, mechanical or otherwise, do it very well? One thing is for certain when looking through any engineering advert – it's jolly difficult for anyone not in the business to understand. Engineering is a huge subject and its adverts tend to be rather specific. You very rarely see an ad that says 'John Smith – General Engineer – no project too large or small – bridges, spacecraft and fibre optic network relays a speciality'. Perhaps it's just that there is so much engineering now that no one can be a generalist. But this certainly wasn't always the case and if we head back some 400 years we can find what is (in my humble opinion) without doubt the best bit of advertising for a general engineer ever undertaken and a book that no engineer should be without. Welcome to the work of Agostino Ramelli and his magnificent, and magnificently monikered, magnum opus, *The Diverse and Artifactitious Machines of Captain Agostino Ramelli*.

Ramelli was born in 1531 on the border of Switzerland and the Duchy of Milan and in 1572, as an apprentice to the Marquis of Marignano, he first appeared as a military engineer at the siege of La Rochelle in the service of Henri, Duke of Anjou. Anjou, later King of France, would go on to employ Ramelli for many years although, as an Italian, he seems to have gained little credit at the French court. Perhaps this was what inspired him to produce his great work in 1588, the year of the Spanish Armada.

It would be a crime not to own a book with a title as good as that and the contents themselves do not disappoint. Agostino Ramelli's book contains 200 lavish illustrations of the extraordinary machines he had invented (and sometimes 'borrowed' or adapted) and forms a sort of exposition of the early engineer's art. *The Diverse and Artifactitious Machines* is perhaps the finest example of a whole new genre of literature appearing in the sixteenth century known as the *Theatrum Machinarum* or 'Theatre of Machines'. It was the engineering equivalent of Chippendale's directory of furniture, offering page after page of mouth-watering designs to appeal to every despot, benevolent dictator and tyrant who had the money to employ Ramelli and his ilk.

For the more belligerent there were huge catapults, machines for filling in castle ditches, devices for scaling sheer walls, bolt cutters for snipping through portcullises like a knife through butter and portable bridges for the army on the move. These were the stock in trade of the sixteenth-century engineer who, as a latter-day Archimedes, could find most ready employment in building machines of war rather than pondering mathematical problems. For those whose martial ambition stopped short of all-out siege warfare there were also tools for removing doors from their hinges or tearing away the bolts that held them shut, which might have appealed more to the jealous lover or cat-burglar.

But Ramelli was no one-trick wonder. He also wanted to appeal to the less warlike leaders of his era. His near contemporary Georg Agricola (1494–1555) had enjoyed great success with his *De Re Metallica*, the finest book on mining and metallurgy since Pliny's *Historia Naturalis*. This was also illustrated with numerous woodcuts showing not only mining techniques but cutaways of the new machinery that could be employed to make it more productive. The idea of digging wealth out of your land had always appealed to European rulers and Agricola's plans for improving and partly mechanising this process were a revelation to them. So Ramelli followed suit and interleaved with the machines for lifting artillery into mountainous places we find intricate illustrations of water pumps, earth excavators and cofferdams, some of which look remarkably similar to some of Leonardo da Vinci's sketches, suggesting that Ramelli had at some point had a peek in the great man's notebook.

Nor were the agricultural sciences ignored. Several cutaway diagrams show improvements for the treadmills and windmills for grinding grain, irrigation equipment and pumping apparatus. There was even room to deal with one of the more civilised pursuits of a Renaissance man or woman – learning. One of Ramelli's more unusual but strangely prescient machines is a device that looks like a waterwheel but with the vanes replaced by reading desks. Each desk is attached to an epicyclic gearing system, meaning that as the wheel is turned, each book remains on its desk without tumbling off. The idea behind the contraption was to allow the avid researcher – and such an idea could only have come from a mind that ranged as far as Ramelli's – to read several books at once, finding a passage in one and then rotating the wheel to another to cross-reference. It was a cumbersome contrivance, whose epicyclic gearing was more for show than function – the device could equally well operate using weights allowing gravity to keep each desk upright – but, in its own peculiar way, it marked a revolution. The ability to read many texts at once and jump between them was a relative novelty and Ramelli's machine has been claimed an even more distant predecessor of hypertext than Vannevar Bush's Memex machine.

Rarely has one engineer covered so wide a field with such aplomb. If there was a machine that Ramelli couldn't build he certainly wasn't telling, indeed one of his engravings doesn't even stop short of the impossible, showing what appears to be an overbalance wheel of the type used in putative perpetual motion machines. But the big question is, did all this advertising work? In the short term the answer is a resounding yes. Ramelli's work became a best seller, a required volume for the library of every Renaissance scholar, soldier and aristocrat. By 1604 Ramelli is referred to as 'Grand Architect to the King', the king being Henry IV of France, but that is the last we hear of him. No date for his death has been discovered. His work lived on however and even reached China in the hands of Jesuit missionaries, his drawings finding their way into Wang Cheng's 1627 *Qi Qi Tu Shuo* (Diagrams and Explanations of Wonderful Machines).

Sadly, in this work, many of Ramelli's machines were misinterpreted by engravers ignorant of the intricacies of his devices; indeed his whole reputation has become somewhat mangled in more recent centuries. Today the Theatres of Machines are looked on as fanciful asides where they should take their place as milestones in the history of technology. Ramelli was the engineer every child wants to be – the inventor, the maker, the improver – and no home should be without his diverse and artifactitious machines.

Episode Five – Return of the Mechanical Turk

Cheating is generally frowned upon in mechanical engineering, although it can be a sore temptation. There is a great satisfaction to be had in creating a machine that can do a job formerly the sole preserve of a human – and sometimes a great deal of fame and fortune associated with it – so perhaps it's hardly surprising that some engineers choose to 'cheat'.

Baron Wolfgang von Kempelen was perhaps the most famous amongst these. The whole affair started around 1769 when Kempelen was visiting the Empress Maria Theresa of Austria at the Schönbrunn Palace. As was usual in palaces there was some entertainment put on, in this case an illusionist who rather impressed those gathered. All except Kempelen that is. For some reason he decided to announce that the illusions were mere trifles and that he was prepared to return to court with something a thousand times better. Boasting to empresses is a dangerous business as they generally take you up on your boast and Maria Theresa was no different. But to Kempelen's great credit he did return, and with something that would have knocked the empress's socks off, had she been wearing any.

It was just a year later that Kempelen unveiled his mechanical wonder. It consisted of a life size model of the head and upper body of a bearded figure, dressed in Ottoman robes and a turban with a long Turkish smoking pipe in its left hand. The figure was seated at a square box on which was placed a chess set. This was all rather lovely but hardly anything wondrous. What did surprise the audience is when Kempelen announced that this manikin was actually a thinking machine and that it would take on all-comers at chess.

Of course the first response to this was suspicion, and so to allay this Kempelen, always the showman, would begin by removing certain panels on the machine and asking the doubters to come and inspect. As they peered into the gloomy interior they could see a complex series of clockwork cogs and gearchains whirring away, proving that the machine was just that – a machine. By removing the front and back panels Kempelen could even let his audience look right through from one end to the other to prove there was no one hiding inside. Now the show could really begin.

Firstly Kempelen would challenge anyone in the room, no matter how proficient at chess, to play the Turk. Now the Turk would suddenly become animated. Its left hand would rise from the cushion on which it normally rested, lift a white piece and make its move. If it put its opponent's king in check it would gravely nod its head three times (twice if it was the queen). It could even apparently chat during games, using a pegboard to indicate answers to any question the challenger cared to ask.

As the match proceeded Kempelen would keep the tension up by peering into a mysterious small box he had placed on the machine as though something inside controlled the game. He would also move around the audience asking them to come forward and inspect the workings, showing the whirring clockwork panels that allegedly were the Turk's 'brain'. And the Turk certainly seemed to be clever. Games were usually short, averaging around half an hour, and the Turk almost always won. After each victory Kempelen would announce that the machine would now prove its prowess by performing the 'Knight's Tour' – a complex puzzle requiring a single knight to cover every square on the board once without repetition.

It's fair to say that the folk gathered at the Schönbrunn Palace were rather impressed, as were the thousands who flocked to see the Turk on its subsequent European tour. Playing the Turk became something of a rite of passage for those who considered themselves great chess players. Faced with the eerie machine which appeared to effortlessly play quite brilliantly whilst answering questions on any topic they chose, many floundered. The best player in France, François-André Philidor, did beat it although he later admitted it was the most exhausting game of his life. After Kempelen's death its new owner arranged matches with Benjamin Franklin and even Napoleon Bonaparte – both of whom it defeated, perhaps helped by its new voice box which could put the greatest minds off their game when it darkly intoned 'Échec'.

But before we get as carried away as late eighteenth-century society there is one fly in the ointment. Very soon a niggling suspicion began to emerge amongst those of a more sceptical bent. For any machine to behave so much like a human wasn't it most likely that it was, in fact, er, human? Well the answer was yes. Whilst everyone wanted to believe that an engineer could make a machine play chess it was simply a hoax.

Inside the 'machine' the key to its chess prowess was that greatest machine of all – a human (known as 'the Director'). This poor soul sat cramped on a moveable seat which could slide out of sight behind a panel of fake clockwork machinery when Kempelen opened one of the inspection doors in the device. The 'Director' followed play by watching the movement of strong magnets attached to the underside of the pieces on the board above. These positions he transferred onto a pegboard where he decided his next move. When he was happy he used a pantograph needle to make his move on the pegboard. This in turn, through a series of connectors, moved the Turk's arm across the board, making its play on the table above. The only light he had to work with came from a candle under a small chimney which vented out of the top of the Turk's turban.

The mechanical Turk was finally destroyed by fire in 1854 and shortly after that the son of the last owner published a series of articles in *The Chess Monthly* revealing the secret. And so the most famous (if slightly dishonest) engineering feat of its time passed into history. By now Kempelen had been dead for fifty years but the epitaph on his tomb '*Non omnis moriar* ...' ('I will not die completely') was not yet redundant. So it was that in 2005 a certain online book shop announced a new computing marvel. A software suite that, at first glance, allowed complex human problems – such as choosing the 'best' photograph of a product – to be performed by computer. In fact the Mechanical Turk, as Amazon called their system in homage to Kempelen's creation, is an automatic system that simply mediates between humans, just as Kempelen's machine did. This Turk is not an attempt to fool the public but simply a computerised way of linking those with problems that even modern computers can't solve with those humans willing to solve them – a marketplace for old-fashioned brain power. And so, despite our advances in technology, we still, it seems, live in an age of mechanical Turks.

Episode Six – The Robots are Coming!

Those of you who haven't been to South Africa might be alarmed at the idea of turning left at the robots; indeed the image conjured is an uneasy one – of street corners inhabited by listless mechanical men brimming with technology but with so little purpose that they can be used as landmarks when giving directions. Of course South African robots are not mechanoids but simply the local term for traffic lights, which came as something of a disappointment to me I must admit. Not that the invention of the traffic light should be dealt with too, er, lightly.

The origin of the traffic light is a splendid story of great British inventiveness followed rather rapidly by a sudden and somewhat explosive reverse. The first true traffic light was produced, rather counter-intuitively, before there were any cars on the road. The man responsible was John Peake Knight, a railway engineer who specialised in designing signalling systems for the burgeoning railways. The signals of his day consisted of a semaphore arm, which was either in the horizontal position or at 30 degrees, to indicate whether a train could pass during the daytime. The semaphore arm also obscured or revealed either a green or red lamp behind it, depending on its position, to indicate whether the track was clear to pass at night or in obscuring conditions when the arm itself might not be fully visible.

The problem that the good burghers of Westminster were facing in 1868 however had nothing to do with trains. Their problem was the huge increase in horse-drawn traffic streaming over Westminster Bridge and what happened to it when it met the thousands of pedestrians thronging the precincts of the Houses of Parliament. So J. P. Knight was approached

13

to find a solution, which he duly did, installing the very first traffic lights at the intersection of Great George Street and Bridge Street on 10 December that very same year.

Not surprisingly his system was effectively the same as a railway one. At its centre was a semaphore arm and a revolving gas lantern with a red and green light which could be operated by a policeman to stop the road traffic to allow pedestrians to cross. The system was a triumph, hailed by everyone except the poor policeman who had to stand next to it all day operating it. In fact just under a month later the policeman on duty had even more cause to rue his luck when, in mid-operation, the whole device exploded, injuring him.

And that was the end of the traffic light. Well not quite. Move the story on another forty or so years and across the Atlantic and engineers were again thinking about the knotty problem of traffic control. Cars had been introduced in the USA in the late 1890s and their relatively high speeds and unfortunate conjunctions with horses, people and occasionally each other meant some form of prioritising was necessary. Perhaps the first man to think about what this should actually be was Earnest Sirrine of Chicago who filed a patent in 1910 for what he claimed was the world's first automated stop–go system using non-illuminated boards with the words STOP and PROCEED written on them. This was all well and good but the lack of lighting meant that their use at night was somewhat limited. Fortunately another recent invention was galloping to the rescue. Electricity grids were spreading across American cities and electric light looked like the perfect, non-explosive, way to ensure the stop and go signs were always seen. Or so thought Lester Wire, a policeman from Salt Lake City who set up the first electric red–green traffic light system there in 1912. Sadly Wire forgot to patent his device, something that James Hoge did not forget to do in the very next year. His lights were installed in Cleveland Ohio in 1914 and added the novel feature of a buzzer to help those who had either nodded off at the lights or, through colour-blindness (or blindness) weren't quite sure what colour they were anyway.

Which brings the story to its final dénouement. The problem with all these systems was that they consisted of just two lights. As red–green colour-blindness is relatively common (affecting 7–10 per cent of men), this was problematic, as was the fact that sudden light changes were hard to react to quickly enough – particularly as traffic speeds increased. The answer came from another policeman – officer William Potts of Detroit. Potts was concerned about the almost impossible task of co-ordinating policemen on a four-way junction to all change their lights at the same time. The answer he found was so simple and effective that it is, fundamentally, the same that we use today. Potts added an amber light (as he had seen on the railroad) and also a timer to co-ordinate a four-way set of lights, cobbled together from a $37 flashing unit used to illuminate signs. In October 1920 the world's first fully automatic, three-light electrical traffic-light system was installed at the intersection of Woodward and Michigan Avenues in Detroit. Potts, like his police colleague Wire, never patented his device, perhaps thinking it just a public service, and it was left to the traffic-light manufacturers to fight it out in court over who should profit from his discovery. He died in obscurity, unrecognised outside his home city. But you can still find officer Potts' robot standing duty at intersections around the world.

Episode Seven – The Forecast Factory

Nothing follows prediction of a blazing summer with more certainty than a complete wash-out. Many of my last few summers have been characterised by building sandcastles in the rain, digging potatoes in the rain and, frankly, doing everything in the rain. My only consolation is teasing meteorologists who have had an even worse time of it, partly because everyone seems to think the weather is their fault, but mainly because predicting the very unsettled weather patterns we've been having is rather tricky and, when they get it wrong, we blame them for that too.

But the story of the development of modern weather prediction and the extraordinary machine first envisaged for doing it is actually a rather heroic tale. Well into the twentieth century most forecasting was done on the 'seen that before' principle. Developing weather patterns would be compared with previously recorded patterns and, if the two looked similar, then the forecast was that a similar sort of thing would happen. Which it sometimes did and sometimes didn't. However, working in the meteorological office in 1913 at the remote Eskdalemuir Observatory was a scientist with a brilliant new idea. It occurred to Lewis Fry Richardson that weather could be predicted by using a mathematical technique he'd invented whilst studying peat, which he'd published just three years before in his catchily titled *The approximate solution by finite differences of physical problems involving differential equations.*

Now the run-up to the First World War was not a great time to be thinking such thoughts, particularly when you're tucked away in Eskdalemuir and certainly not when you're a Quaker and hence an ardent pacifist. With the outbreak of war Richardson became a conscientious objector and in 1916, like many Quakers, took on a non-combative role in the Friends' Ambulance Unit attached to the 16th French Infantry Division. It was an extremely dangerous job: indeed the most decorated NCO of the whole war, William Harold Coltman VC, DCM & Bar, MM & Bar, was a pacifist stretcher bearer who won his medals without ever firing a shot.

While Richardson was ducking shells he decided to test whether the weather on one particular day could be predicted mathematically, so took a day in the past for which the weather was of course known to see if he could predict what would happen just from the starting set of conditions. The day he chose was 20 May 1910 which had been particularly studied thanks to an experiment in which hundreds of weather balloons had been launched across central Europe to record weather patterns at precisely 7 a.m. that day. With just a pencil, paper and slide rule, Richardson then used his numerical process to calculate the weather over one location in southern Germany for the following six hours. The result was a disaster, his calculations predicting a world record increase in pressure over the day when actually pressure remained fairly level. It was, admittedly, a bad start but Richardson persevered.

Richardson's work in the trenches was rewarded after the war with indifference and hostility, his pacifism effectively barring him from most academic posts. Initially he returned to the Met Office but when that merged with the Air Ministry in 1920, he felt conscience-bound to resign as he could not work for a military organisation. A lesser man might have been put off by this, but not him and in 1922, now working on the very outskirts of the academic world, he published his great work on weather forecasting by mathematical modelling: *Weather Prediction by Numerical Process*.

In this he imagined a wonderful machine. A huge, spherical room with the map of the globe painted on the interior around which would be tiered galleries filled with desks where 64,000 people would sit. Each one of these people would be a 'computer' responsible for the equation, or part equation, modelling the weather on the small part of the globe behind where they sat. Around each desk small signs would flash the results of each individual's calculation to those around them while in the centre, on a tall podium, the 'weather conductor', a person responsible for keeping the calculations of all 64,000 individuals running in perfect harmony, would stand. If some areas fell behind, the conductor would shine a blue light on them to tell them to speed up, if others ran ahead, a 'beam of rosy light' would be shone in their direction to tell them to slow down. Around the conductor four clerks would dispatch the results by pneumatic tube to a telegraph office where they could then be radioed to the waiting world.

Richardson's weather machine was of course just a dream in 1922. There were no machines that could make so many calculations in tandem and no one seriously considered putting 64,000 mathematicians in a spherical room for any length of time. But the idea was prescient. Richardson had realised that by modelling weather patterns and running those models, weather could be predicted mathematically far more accurately than by simply using precedent. By the time Richardson died in 1956 the electronic computers capable of making his dream a reality were just coming into existence and his numerical process soon became the single most powerful tool in meteorology. It is still not perfect but before we complain about it we should remember that without this gentle, widely overlooked genius we'd still just be guessing that tomorrow's weather might be a bit like today's – and I certainly hope that's not true.

Episode Eight – Engineering is not always about 'stuff'

For many people engineering is thought to be about making stuff – real, solid, physical things – and to prove them wrong I'm going to start with a 'mystery quote':

The historian, with a vast chronological account of a people, parallels it with a skip trail which stops only on the salient items, and can follow at any time contemporary trails which lead him all over civilization at a particular epoch. There is a new profession of trail blazers, those who find delight in the task of establishing useful trails through the enormous mass of the common record.

Sounds exhausting doesn't it, all that trailblazing through enormous masses, and the device that historians, like myself, were envisaged as riding on this epic journey was even more alarming. But what makes this prediction most frightening of all, is that it has all come true. The quote above comes from an article entitled 'As We May Think' by Vannevar Bush which appeared in the July 1945 edition of *The Atlantic Monthly*. He was envisaging a machine he called the 'Memex' or 'memory extender'. This being 1945 the Memex was not your average laptop, however, but an electronically controlled desk which sounds like it might have escaped from a Terry Gilliam film. The device allowed the operator to scan through books and articles using two microfilm readers and create links between items, so rather than just reading book after book, the user could create trails of information through many books, articles, photographs and other data. These trails were given a code which the owner recorded in their code book so when a useful trail needed to be recovered it could be called up just by punching in some numbers on a keypad. New information could also be added to the Memex thanks to a photographic plate, rather like a modern scanner, and manual notes could be appended to parts of the trail. The desk in its finest form, as described in *Life* magazine on 10 September 1945, also included a head-mounted camera allowing operators to record experiments as they performed them, a voice recognition typewriter and speech synthesis.

But the Memex was sadly never to become a regular item of furniture, indeed the idea was far ahead of the technology needed to bring it to life. What it did do however was suggest a whole new way of looking at information, one which concentrated on the links between pieces of information as much as on the information itself. This provided the direct inspiration for the development of hypertext, the ubiquitous 'language' of the world-wide web, by Douglas Engelbart and Ted Nelson, which allows what was once the static immovable type of book pages to become a dynamic system allowing users to explore the meaning of text by taking journeys through the links embedded within it.

Astonishingly, and unlike so many predictions of future technology, Vannevar's proved rather accurate. As a historian today I don't sit down with every book on a subject and read all of them; in fact I don't have to deal with physical volumes of material very much at all. Instead my research is a series of 'skip trails' through the ever-expanding ocean of data contained on that great Memex in the aether – the world-wide web – with the occasional stopover in real libraries to make sure I haven't been overly beguiled by Wikipedia. In fact Vannevar predicted Wikipedia too, in the same article, when he wrote:

Wholly new forms of encyclopedias will appear, ready made with a mesh of associative trails running through them, ready to be dropped into the Memex and there amplified.

Of course there were attempts to deal with the huge corpus of material available in every subject before the Memex. Monks have glossed books, indexers compiled lists, and clerks carefully filled in index cards for many centuries before microfilm or magnetic storage came along. But they all suffered from the problems of the physical storage of media and accessing it amongst hundreds or thousands of volumes kept in different geographical locations. They also used proprietary cataloguing systems which lacked any linkage between them – all things that the Memex might, to some degree, also be accused of.

Nor was the Memex true hypertext, just a way of recording paths through data, and it didn't actually make individual items in that data active links in themselves. It didn't have a search engine, which would make finding new information sources quite tricky and there was no standard metadata to tag items with so that they could be easily searched. But it was 1945 after all and Bush had actually had the idea in 1939, before the Second World War so rudely interrupted his thoughts.

So what exactly is there to celebrate about a man who invented a machine that was never built, might never have worked and which lacked the tools that later generations have found essential for undertaking the sort of work he envisaged his machine doing? This is where Vannevar Bush proves that engineering isn't all about physical 'stuff'. What he engineered wasn't a thing but an idea, and, in my opinion, one of the greatest ideas of the century. Bush imagined new ways of navigating through data, new tools for exploring information and, in an era long before the personal computer, he imagined what those journeys might be and who might take them – not just scientists and engineers but even historians. He imagined a time when no profession was untouched by what we might call information technology:

The lawyer has at his touch the associated opinions and decisions of his whole experience, and of the experience of friends and authorities. The patent attorney has on call the millions of issued patents, with familiar trails to every point of his client's interest. The physician, puzzled by a patient's reactions, strikes the trail established in studying an earlier similar case, and runs rapidly through analogous case histories, with side references to the classics for the pertinent anatomy and histology ...

Bush may have been talking about an unrealised dream, but there can be few more accurate predictions from 1945 of the world of today and so, as a historian, I think 'As We May Think' must go down as one of the greatest engineering feats of the twentieth century – without not so much as a spanner being lifted.

III

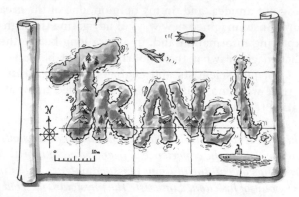

In which we explore the desire of engineers to get us to go higher and faster and deeper – whether we want to or not.

Episode Nine – Up, Up and Away

As I was sitting in the departures lounge at Bristol Airport in the summer of 2009, I was reminded that the world's very first commercial airline went into business exactly 100 years before, and to be honest it wasn't a thought to fill any prospective passenger with confidence.

The catchily named Deutsche Luftschiffahrts-Aktiengesellschaft (German Airship Travel Corporation), known as DELAG, was an unusual airline, the brainchild of the equally catchily named Ferdinand Adolf August Heinrich Graf von Zeppelin. Zeppelin had studied engineering at Tübingen but found his calling in America during the Civil War when another German, John Steiner, offered him a flight in one of the tethered balloons that the Union Army had been using for reconnaissance.

From this first flight Zeppelin was inspired to create a whole new type of transport. He imagined a balloon not tethered or subject to the vagaries of the wind but steerable and powered, like an ocean liner in the sky. Except Zeppelin wasn't really thinking about passengers at this time – he saw the airship as a weapon of war.

By 1890 when he retired from the military, he had already begun sketching out plans for a rigid-framed craft filled with individual gas cells. Teaming up with balloon engineer Theodore Kober, construction work on Zeppelin's first airship, LZ-1, finally began in 1898. Two years later, inside a floating hangar on Lake Constance (the unusual location allowing the whole hangar to be turned into the wind when the airship entered or left its moorings), the strange vessel was first filled with the hydrogen gas that was going to give airships such a lively later history.

It was not the most auspicious start in truth. Zeppelin's construction superintendent had, at the last minute, refused to board his own craft, claiming that he didn't have enough insurance cover. He was dismissed shortly after. But he need not have feared, as the almost 130-metre-long craft with its two metal gondolas swung beneath it, emerged from its cocoon and floated gracefully down the lake for 18 minutes. It was an undoubted success, if a qualified one. The tubular-steel frame was seen to be sagging alarmingly at each end, whilst the lack of fins made the craft rather unstable to fly. Power was provided by two hopelessly inadequate 14-horse-power Daimler engines whilst pitch had to be controlled by a sliding weight suspended on ropes under the airbag. This got jammed, which didn't really help things. The military observers at the first three flights were less than impressed and refused funding. Zeppelin was forced to dismantle his machine and go back to the drawing board.

But if the military had been unimpressed, the public was strangely taken with the airship and five years later Zeppelin was again on the waters of Lake Constance with his second attempt, LZ-2, this time partly-funded by a public lottery. Sadly it suffered engine failure on its first flight, was forced to make an emergency landing and was later destroyed in a storm. But LZ-2 was an improvement, if a short-lived one. Zeppelin's new chief engineer Ludwig Dürr had replaced the tubular girders with much stronger triangular ones and the engines now boasted 80 horse power each – or they would have done if they'd worked.

Zeppelin realised that he had found a unique engineering talent in Dürr and his ideas were put to work in the rather unimaginatively named LZ-3 and LZ-4. These sported stabilisation fins and elevators and proved so safe that the King of Württemberg and his wife were persuaded to take a trip. The German government now woke up and offered to fund further development but only if the new airships could make a daring 24-hour flight. The challenge began on 4 August 1908 but sadly didn't end 24 hours later. Instead the LZ-4 was forced down by a storm at Echterdingen where it dragged its moorings and exploded.

Now you might think this would put a damper on things, but instead the German public took to the idea of air travel even more passionately, and the crash became somewhat contrarily known as 'the miracle of

Echterdingen'. So famous indeed did Zeppelin become that he could set up a proper airship company, the Luftschiffbau Zeppelin.

The question was, who was going to buy his airships? The military were still not all that keen and so Zeppelin and his business manager decided to form a company providing commercial flights in airships. In 1909, DELAG was formed as the world's first commercial airline, and a year later, on 19 June, commercial flights began on the LZ-7 'Deutschland' carrying passengers from Frankfurt to Düsseldorf and Baden-Baden. Sadly this crashed just nine days after its maiden voyage. LZ-8 'Deutschland II' proved not much more successful, being smashed against its own hanger a month and a half after launch. Clearly an innovation was needed to calm passengers' nerves. It arrived in 1912 on the LZ-10 'Schwaben' in the form of Heinrich Kubis, the world's first onboard air steward. Sadly the LZ-10 also blew up a year after launch.

You can probably now see why airlines don't put up illustrated histories of the first airline in departure lounges. But DELAG was by no means as fatal as it seemed. Despite the explosions, no one was killed or even hurt and between 1910 and 1914 over 34,000 passengers were carried on over 1,500 flights, without a single injury, making them one of the safest airlines in history.

Episode Ten – By Way of a Small Demonstration

Picture the scene – you stand outside the boardroom of a major international company, the patent for your brilliant invention clutched firmly in your hands. You know it's a great idea, the patent examiner knew it was a great idea, even your cat seemed interested in it, but what will the ladies and gentlemen in sharp suits on the other side of the door think? The time has come, you walk in and for the next twenty minutes you give yourself body and soul to explaining why your invention will change the world. In return you're met with a spectacular wave of indifference and boredom from the serried ranks of executives too set in their ways to ever contemplate anything so radical for fear it might affect their share options.

So what went wrong? Well in the first place you can comfort yourself with the thought that, historically, this has been a very familiar scene. But that's enough self-pity, for some of those who found themselves in your shoes fought back and won and did so by means of a small demonstration. Take Charles Parsons.

Charles Parsons shouldn't really have had much trouble getting support from the establishment. He was, after all, the sixth son of the 3rd

Earl of Rosse, sometime proud possessor of the world's largest telescope, so he had some excellent contacts. He had also received a thorough scientific education thanks to his father who had employed the Astronomer Royal as his tutor. Nor was his father simply interested in giving his sons a theoretical grounding, encouraging them instead to get their hands dirty in the workshops and laboratories of their home at Birr Castle in Ireland. So the adult Charles Parsons that emerged in 1877 after spells at Trinity College Dublin and St John's College Cambridge (where he graduated 11th wrangler in mathematics) should have found every door open to him (even allowing for the unfortunate incident when his cousin Lady Bangor fell off his home-made steam car and was killed).

A great engineer however must fight for recognition – even if they are the mathematically talented, dangerous-car-building son of an earl. The young Parsons undertook a string of apprenticeships to learn the practical elements of engineering before settling down in 1884 to solve the tricky problem of electricity generation as a partner in a Gateshead firm. The problem was simple – the dynamos used to generate electricity were powered by low-velocity reciprocating engines attached by a belt drive running at around 1,000 revolutions per minute, leading to very low efficiencies and very low generating pressure. Parsons' simple but brilliant idea was to do away with nearly all the machinery and rather than condense and evaporate the steam through pistons and chambers, just let the steam race directly, in a series of controlled stages, through a disc of vanes, around a central spindle which was also the driveshaft of the dynamo. And so the steam turbine was born.

His 18,000 revolutions per minute turbo-dynamo was an instant hit, particularly aboard ships where small and efficient generators were needed for lighting. But Parsons realised that his turbine was more than just a neat way of generating electricity – it was a whole new way of delivering power – and ships seemed to him to be the ideal way of exploiting this. His idea, patented in 1884, was to power not just the lighting but the driveshaft of the main screw using a turbine instead of a traditional steam engine. Increased mechanical efficiency would mean a reduction in fuel consumptions and increased revolutions per minute would translate as faster speeds on the water. Surely the Admiralty would jump at this? At a time when Britain was the world's leading naval power, the opportunity to build even faster, more efficient ships using proven technology promoted by a man who was both widely acclaimed as an engineering genius and the son of an earl to boot was surely too good to miss. And so Charles Parsons found himself in that 'standing outside the door clutching a patent' situation familiar to many engineers. And he explained to the Admiralty his wonderful idea and showed them the drawings and they no doubt muttered amongst themselves about what the son of an aristocratic astronomer was doing trying to tell them their business and, in short, they completely ignored him. Why did the British navy need steam turbines when they already had destroyers capable of the dizzying speed of 28 knots?

A lesser man may have been disheartened but Charles Parsons was not one to give up so he decided upon a little demonstration. The year 1897 was to be the Diamond Jubilee of Queen Victoria – a momentous occasion to be marked by a full naval review at Spithead at which the Queen could gaze on with satisfaction at the huge state-of-the-art fleet which ensured that Britannia ruled the waves. And Parsons intended to be there, uninvited. And he would not come alone. As the 26 June 1897 dawned and the dignitaries began arriving and the warships began forming up in lines off Spithead, another much smaller ship was creeping out of harbour unannounced. She was Parsons' new vessel – a 100 ft, 44 ton craft powered by three screws running off three turbines and called, appropriately, *Turbinia*. She had had a hard birth. Running ship's propellers at high speed off turbines had created such cavitation that the project was nearly scuppered before it could get started, but Parsons had spent months redesigning the propellers, their housings and the hull as well as studying models to see how the bow wave and wake of a high-speed ship would affect its performance. Now he was ready – 165 ships lay at anchor off the shore and the Prince of Wales (representing Queen Victoria who was too frail to attend), Prince Henry of Prussia and the whole Admiralty board climbed the specially built stage to watch proceedings.

Then as the National Anthem struck up and the assembled dignitaries rose to their feet, Parsons raised a bright red pennant on *Turbinia*, opened the throttle and out she shot at 34 knots between two lines of lumbering British warships, right past the royal party and just narrowly avoiding a French yacht. The Admiralty was furious at being upstaged and signals were quickly sent to intercept the interloper. Patrol boats swung into action and the chase was on. But it was an unfair match. *Turbinia* easily outran the slow, steam-powered navy vessels – she was, after all, the fastest boat in the world.

By that evening the assembled naval dignitaries from across the world were not celebrating Britain's steam-powered fleet, but the death of that technology and the birth of a new one. Within fifteen years all new British warships and Atlantic liners were turbine powered. Thanks to one audacious demonstration a new age of steam had been born.

Episode Eleven – On the Buses

We live in straitened times, so the newspapers seem to delight in telling us, and it is perhaps only right to make some belt-tightening resolutions. The chauffeur will have to go, cab fares are now an unaffordable luxury and so it's time to get back on the buses.

That I can get back on the buses is all thanks to an event that happened over 180 years ago, courtesy of another man who regularly found himself in a spot of financial bother – George Shillibeer. George was an engineer and entrepreneur of a practical bent. Born in 1797, he had spent a brief spell in the navy during the Napoleonic Wars before deciding on the safer option of apprenticing himself as a coach builder to Hatchetts in Long Acre. In 1825 George went to Paris where he worked on the design and construction of a new type of coach intended to carry large numbers of people (based on a similar service already running in Nantes). This gave him an idea, and the following year he was back in London, in business with John Cavill, a coach builder and livery-stable keeper in Bloomsbury.

George's big idea was no less than the creation of the world's first urban mass-transit system – or buses as we might call them. His first commission came from the Quaker College in Stoke Newington and was to design and build a 25-seat carriage to take young ladies from the Quaker Meeting House in Gracechurch Street to and from their school. The result was the first school bus, which delighted one visitor, Joseph Pease, so much that he wrote a poem about it. It would be churlish not to reproduce it here being as it is the first poem ever written about a bus:

The straight path of Truth the dear Girls keep their feet in
And ah! it would do your heart good Cousin Anne
To see them arriving at Gracechurch Street Meeting
All snugly packed up, 25 in a van.

George Shillibeer had a much bigger idea for his machines, however, and was thinking of setting up in business against the hackney carriages and stagecoaches. These forms of transport were exclusive and expensive and hence simply unavailable to most of the population of London. What if he built wide, stable carriages to take twenty or so people for a fixed fee on a set route, allowing them to hop on and off wherever they please? George initially thought of calling such a machine 'The Economist' but thought better of it and settled on 'Omnibus' – something for which we should all be thankful as no one wants to travel atop a 97 horse-power economist.

His big break then came in 1829 when the Quakers no longer needed their bus following the opening of a Friends House nearer the school. And so the girls of Fleetwood House had the privilege of watching a little piece of history as their bus was repainted in the schoolyard with the words 'omnibus' and 'Paddington to the Bank'. The London bus had been born and on 4 July of that year it made its first fare-paying journey.

And so George Shillibeer's fortune was made? Not quite. No engineer's life is that simple. George's three-horse omnibuses were soon up against stiff competition. For a start the Hackney carriages still held a monopoly in the centre of town so he couldn't go there, whilst outside this area smaller buses utilising just two horses were soon more frequent, cheaper, faster, more manoeuvrable and paid a lower mileage duty. On 4 March 1831 George was declared bankrupt. His saviour, oddly enough, came in the form of his competitors. So good was Shillibeer's idea that London was soon heaving with competing buses, engaged in sometimes lethal races to get passengers, and so, six months after his bankruptcy, the operators met to regulate their business, appointing George as chairman. The number of buses was reduced, timetables were introduced to prevent racing, and routes were allocated to operators which could be bought and sold, allowing larger companies to grow.

With the ending of the Hackney cab monopoly in 1832 buses were finally allowed into central London but George's three-horse carriages proved too wide for the narrow streets. Never easily dismayed, he invented the long-distance bus route – initially London to Brighton aboard the unusually named 'New Improved Diligence'. But this too was soon outper-formed by the new railways and again George became bankrupt and had to escape to France to avoid his creditors. On his return he spent time in the Fleet Prison for debt before throwing in his hand with the opposition and joining the London & Southampton Railway Company. Sadly, an unfortunate indecent with 130 gallons of brandy that had somehow escaped the notice of customs officials brought this job to an end and he was sent to prison again.

From prison George tried to persuade the government to recognise his work in introducing the bus but his entreaties fell on deaf ears. In desperation the newly released George turned again to Paris for inspiration and 'borrowed' from there the idea of the funeral omnibus, combining a hearse with a bus for mourners to reduce the cost of funerals. His last years were hence spent more quietly and profitably as an undertaker. He died on 22 August 1866 largely unknown to the thousands of people who by then travelled in London buses every day.

Episode Twelve – Flying Trains

We've already talked about Zeppelins, in particular those that went up, up and away and then occasionally burst into flames, but what about those that ran on rails? I say 'those', there was only really one, in Germany at least, but for a moment in the 1930s it looked like the sort of high-speed train transport that still fills the pages of science fiction was going to get off to a flying start.

The Schienenzeppelin was not the first propeller-driven railcar. Those laurels probably belong to the Soviet Aerowagon which Valerian Abakovsky designed to whisk important Soviet officials around their huge country. Sadly, on 24 July 1921 it whisked all those on board, including Abakovsky, off the tracks and to their untimely deaths, marking the end of a rather promising project. More fortunate was the brilliant Scottish engineer George Bennie who built a prototype track and railcar for his 'Bennie Railplane' at Milngavie near Glasgow, which opened on 8 July 1930. His idea was for a propeller-driven train riding on a gantry high above the original track, carrying passengers quickly and quietly to their destinations, while the slower freight trains lumbered on below. Everyone agreed it was a magnificent idea and Bennie was feted as the next in a long line of those great British engineers who were building the modern world. Showing a depressingly familiar lack of foresight, both government and private backers then failed to invest anything further in the project, leaving Bennie bankrupt and his test track sold off for scrap.

This, however, was not how they were going to treat the future in Germany. They too had a brilliant engineer, in the shape of Franz Krukenberg, originally a ship-building engineer, who had already designed aircraft and airships, although he was an early critic of Zeppelins due to their explosive nature. In 1929 he came up with his own spin on the idea – the Schienenzeppelin (rail zeppelin) – so called simply because it looked a bit like a Zeppelin airship. Built from aircraft aluminium, the light, single-unit railcar would carry forty passengers down the track, pushed along by a rear propeller attached to two (later one) huge BMW aero engines. By early 1930 an enthusiastic German state had the plans on the drawing boards of the designers of the German Imperial Railway company. Just six months later it was on the tracks. With its sleek aerodynamic bullet nose and modernist, stripped-back, Bauhaus interior it looked frankly more advanced than most trains that I at least currently travel on.

And it was fast. On 10 May 1931 the train broke the 200 km/h barrier and just eleven days later set a new world rail-speed record of 230.2 km/h. This wouldn't be beaten until 1954 and is still the record for a petrol-powered train.

So what happened to this future? It won't have escaped your attention, when sitting at faulty signals outside Woking for an hour or so, that we still don't whiz around in high-speed aircraft-style railcars. Well to be fair, the Schienenzeppelin was not a perfect design. The train proved unable to climb steep gradients hence requiring either new tracks to be laid or an additional power-source put in the unit. There was also something of an infrastructure problem – one which has hardly gone away since. As the train was fast and light there was a danger that uneven tracks, designed for heavy, slow trains, might lead to the Schienenzeppelin 'lifting off'. To try to counter this the driveshaft was inclined at a seven-degree angle to provide some down-thrust, but the truth of the matter was that not many tracks were designed for a train quite like this and track laying was a much bigger and more costly undertaking than train building. There were other more prosaic problems too. The rear propeller drive meant that the train couldn't really be attached to other units, hence nullifying one of the great advantages of rail transport – the ability to connect large numbers of wagons or coaches together. Finally there were those spoilsports who pointed out that a huge uncovered propeller, scything through the air at terrifying speeds, made standing idly on platforms a shade more dangerous than it had previously been.

All these things conspired against the futuristic Schienenzeppelin and in 1939, at the outbreak of war, the only prototype was broken up – its parts apparently needed for the war effort. The German Imperial Railway took another tack, developing its own non-propeller-driven railcar, the splendidly named 'Flying Hamburger'. So had Germany's Zeppelin train been a red herring? As is so often the case, much of what Krukenberg learnt by building the Schienenzeppelin was put to use in later high-speed designs such as the TGV, but not everyone at the time was all that unhappy. At the risk of suggesting the Nazi state was at all cynical, to many in the regime the programme served its purpose admirably. The 1919 Treaty of Versailles had forbidden Germany an airforce but who could complain about the development of high-performance aero engines and airframes if they were simply to be used in trains?

Episode Thirteen – Go To Work in a Bean

Henry Ford, founder of the Ford Motor Company and father of the modern assembly line, has been called many things but rarely 'green', yet had he had his way and were it not for the outbreak of the Second World War we might all now be driving around in cars made partly from beans.

Henry Ford had been brought up on a farm so agriculture was not as alien to him as perhaps we might expect of your average industrialist. As an industrialist however he did believe that agricultural output could, with enough thought, be put to industrial use. In particular his interest was piqued by the humble soya bean which to him appeared to be the solution to many of the problems facing America in the Depression. Firstly the plant had nitrogen-fixing roots which he believed could help stabilise and improve the ruined agricultural lands of the 'Dust Bowl'. Secondly the beans themselves were high in oil and oil interests people who make cars. Then there was the fibrous protein left behind after oil extraction and Ford certainly wasn't going to waste that, even if he couldn't initially think what to do with it. Fortunately by the early 1930s he had a research team looking into just that.

The oil proved the easiest to utilise. Early Ford motor cars were painted with seven layers of hand-finished lacquer which was both time-consuming and expensive. Research showed that soya oil mixed with enamel paint provided a quicker, harder and shinier finish without the need for all that polishing. By 1935 some 1,000,000 gallons of soya oil was going into this alone. The oil could also be used to derive glycerine of which 540,000 gallons a year were soon going into his cars' shock-absorbers. Even the Ford foundry had a use for the oil, using 200,000 gallons of the stuff as a binder for the sand in casting moulds. By the late 1930s Ford required 78,000 acres of soya beans a year to make its cars.

Ford was not a man to tolerate waste, however, and all that oil left behind a mountain of fibrous protein. After much experiment it was discovered that this could be mixed with formaldehyde and phenol to make a type of plastic that could be dyed in the production process and brought to a high polish, eliminating the need for painting. Thanks to this, Ford cars soon found themselves with soya glove-compartment doors, gear-stick knobs and electrical buttons. Ford tractors even got soya seats.

If a tractor seat could be made from beans, Ford soon realised that perhaps the rest of his car bodies could also be 'grown'. As he commented in 1934: 'Someday you and I will see the day when auto bodies will be grown down on the farm.' By the late 1930s his soya research team was producing prototype car panels from the material in the hope that plastic panelling would prove more resilient and less likely to dent than sheet steel. It was also noted that a car with plastic panels would be rust-free, a lot lighter and hence far more efficient.

The first soya panel was attached to the boot of one of Ford's own cars in 1940 and in November of that year he arranged one of his famous publicity stunts to announce it to the world. Inviting the press to see his car, the 77 year-old suddenly produced an axe, its blade covered, and proceeded to slam the back of the axehead into the boot. The axe bounced off, the soya-bean lid proving its superiority to steel. He then asked the gathered pressmen if any of them would like to see the result of him hitting the boot of their cars with his axe. Not surprisingly no one volunteered.

The first all-plastic panel car made its debut at another of Ford's great public events, the Dearborn Days Community Festival on 13 August 1941. There is some doubt as to the exact composition of the panels in this car, with sources claiming that the research team used phenol plastic, derived from coal tar, as soya plastic had proved hard to make waterproof. However, plastic, of some form, it certainly was. Ford himself maintained it was a soya car and to prove the huge benefits of the bean he championed he invited the gathered press to share a 14-course meal with him, consisting of only soya-derived dishes.

Strangely it was perhaps that meal that proved Ford's lasting contribution to the soya industry. Ford had asked his chief scientist Robert A. Boyer to experiment with making fabrics from soya fibre, and could even sport a suit made of 30 per cent soya fibre and 70 per cent wool, to match a tie made of the same. Boyer also made his wife a soya coat, and in the process had the idea for 'knitting' textured soya protein into meat substitutes, a food that is still with us today.

The bean car sadly is no longer with us. With the war taking its toll on bean research, the pilot plant built to make soya protein fibre was converted to producing aircraft engines and by the end of the war petrochemical-derived plastics were proving themselves more versatile and much cheaper than soya ones. At Ford's death in 1947 it is estimated that he had spent over $4 million (around $50 million today) on soya research but his 'green' car had never got off the staring line. Instead the result of all that money and work is the humble veggie burger.

Episode Fourteen – Going Down

Now engineers are a practical bunch, so I assume that, from time to time, they've all considered building a submarine. Whilst this is in no way a long-term solution to global warming it might at least give greater minds than mine a bit longer to think about a solution. In the meantime I shall busy myself looking at the still largely unexplored undersea world.

The first surviving record of anyone wanting to know what lay at the bottom of the sea comes from Aristotle who claims to have seen divers placing inverted cauldrons over their heads so they could reach the seabed, although he doesn't say why they wanted to go there. Alexander the Great allegedly had himself lowered to the sea floor in a huge glass bottle and he returned to the surface with a prophecy, claiming:

The world is damned and lost. The large and powerful fish devour the small fry.

Viking sailors had a practical interest, however, in knowing what was on the bottom, as this sometimes came up to meet them rather suddenly in the form of reefs and rocks. To avoid these dangers they invented the sounding weight – a lead cylinder with a hollowed out end filled with something sticky like beeswax. The weight was attached to a line and dropped overboard. When it hit the bottom the rope could then be hauled in, the depth being measured by counting the number of arm-spans of rope pulled up. This 'arm span of an average sailor' became a standardised measurement of six foot, or one fathom as sailors called it. The wax on the end of the weight was useful as, by looking at what stuck to it, you could get an idea of what the bottom was made of. This helped with choosing good anchorages and preventing raiding parties ending up high and dry on the rocks at low tide and hence outstaying their welcome.

Despite the invention of the sounding weight, ships still continued to sink and it was a desire to recover what should have remained on the surface rather than see what dwelt on the bottom that inspired the first true diving machines. One of the first to have a go was Leonardo da Vinci who drew plans for a diving helmet complete with spikes to protect the wearer from the 'monsters of the deep'. Like so many of Leonardo's inventions however, there is no evidence he ever built one.

The first person to think about how to explore the undersea realm properly was the British mathematician William Bourne who in 1578 drew up plans for a submersible enclosed rowing boat. He was not foolish enough to have a go at actually building one however. Unlike Cornelius Jacobzoon van Drebbel, a Dutchman working for the British navy, who was, and is hence usually credited with having invented the submarine. Van Drebbel, who is normally and unaccountably depicted in engravings with a semi-naked woman, had an interesting but ultimately unfulfilling life at various European courts. An engraver by trade, his passion for inventions first brought him to the attention of royalty when he invented a self-winding clock which operated by changes in air pressure. King James I of England thought this was a 'perpetual motion machine' and appointed him 'mechanic to the Prince of Wales', charged with arranging suitably elaborate firework displays and other entertainments for the prince. The Emperor Rudolph II went one stage further and declared it to be magic, appointing Van Drebbel 'Chief Alchemist'. Life at the Prague court was not pleasant however, particularly after Rudolph went completely mad and was deposed and Van Drebbel (for fear of his alchemical powers) was put in prison. On being finally released he returned to England where he swapped alchemy for the less dangerous work of inventing microscopes, devising one of the very first feedback-controlled devices (a chicken incubator whose temperature was controlled by a mercury thermostat), creating a new scarlet dye and building the first submarines.

His early submarine was actually a form of Bourne's underwater rowing-boat, consisting of a greased leather-clad barrel with oars sticking out of either side – the rowlocks being sealed with leather flaps. When submerged, the proud owner of a 'Van Drebbel' then had to row (or rather get his oarsmen to row) with considerable force, not only to move forward but to dive and surface. Tests were undertaken with the sub in the Thames between 1620 and 1624 in three successively larger vessels – the final leviathan having six oars and twelve rowers – and, judging from what was flowing into the Thames at that time, neither the smell nor the view can have been particularly pleasant.

Quite astonishingly, the Van Drebbel submersible did work and managed to operate at depths of up to 15 feet for several hours at a time, cruising from Westminster to Greenwich. So impressed was King James I that he is reported as having taken a trip in one of Van Drebbel's later models, which, if true, would make him the first monarch to have travelled under water (excluding those who were drowned and assuming the Alexander the Great story is just that).

So at last Van Drebbel had a success on his hands? Er, no. The king might have liked the idea but so unimpressed was the British navy that it cancelled the project, turning its back on one of the most important naval vessels of all time – at least for now. This was despite the predictions of the Bishop of Chester – who was a very clever man and the only person ever to be master of colleges at Oxford *and* Cambridge – who pointed out in his 1648 book *Mathematicall Magicke* that:

'Tis private: a man may thus go to any coast in the world invisibly, without discovery or prevented in his journey.

and:

It may be of great advantages against a Navy of enemies, who by this may be undermined in the water and blown up.

But the navy knew best and Drebbel and his great idea were forgotten. After an unsuccessful attempt to blow up La Rochelle harbour on behalf of the British navy, Drebbel fell into disgrace and ended his days as a penniless ale-house keeper. Despite this, there is a crater on the moon named after him. There are no craters named after James I or Rudolph II.

IV

In which we discover that it's not easy to find out who invented what and that when you do there's likely to be a fight.

Episode Fifteen – Bright Sparks

The mastery of electricity is perhaps what marks out modern society more than anything that has gone before but we're not really even sure who discovered it. If you'd asked Edison he would have said that he did, but then he said that about most things. In fact the story begins 2,500 years earlier in Miletus in Greece.

Thales of Miletus is remembered for two things today – believing that everything was made of water and falling into ditches. The former idea, though wrong, is said to be the first step in the history of science, the first time someone tried to explain the world around them without invoking God. The latter, which occurred a lot as Thales liked to look up at the sky whilst walking along, also (according to Plato) had the added benefit of attracting young women, who would help him out of ditches and tease him about his absent-mindedness. This only encouraged Thales to fall into more ditches. But Thales also discovered a third thing – that by rubbing amber on fur, the hairs on the fur could be made to stand on end; indeed if rubbed hard enough a spark could be seen jumping between the two.

Progress after this was frankly slow. It was AD 1600 before anyone decided to give a name to this phenomenon – electricity – ironically enough choosing the Greek work 'elektron' which means 'amber' and which Thales already knew. That man was William Gilbert who was also the first person to realise that compasses point north because the whole earth is a giant magnet – not because there is a large floating magnetic island near the North Pole (as was previously thought). Sadly, just after realising this he died of Bubonic Plague, further slowing the history of electricity.

Otto von Guericke was the next bright spark, inventing the first electrostatic generator in 1660, but he wasn't really sure what to do with it and decided instead to invent the scourge of every school science lesson – 'Magdeburg hemispheres'. Having created electricity it was another 85 years before anyone worked out how to store it – in Leyden Jars, invented by Pieter van Musschenbroek at Leyden University. What he was saving it for however he wasn't quite sure and it was another two years before William Watson realised that what it discharged was an electrical current.

At exactly the same time and on the opposite side of the Atlantic another scientist, freedom fighter and part-time politician called Benjamin Franklin was also experimenting with electrical discharges but on an altogether larger scale. He'd realised that lightning was electricity and proved it by performing the most dangerous experiment in the history of science – flying kites in thunderstorms. Fortunately for Franklin, he understood about insulation and hence never actually waited for his kite to be struck by lightning. The Russian scientist Georg von Richmann was not so lucky. In attempting to copy Franklin's experiment, and perhaps due to an unfortunate flaw in the translation he was using, he attached himself directly to his kite and waited for lightning to strike. It did, and he was spectacularly killed by ball lightning hitting his forehead. In the process he became the first human version of another of Franklin's great inventions – the lightning conductor.

Franklin's work was now inspiring a whole generation (if you'll pardon the pun) of electrical experimenters. In Italy Luigi Galvani, spurred by Richmann's explosive exit from the world, realised that by applying electricity to the nerves of dead frogs he could reanimate them, making them twitch. From this he deduced that electricity was the 'stuff of life' and, more importantly, laid the scientific background for Mary Shelley's great horror novel *Frankenstein*.

But whilst all this twitching was great fun no one yet knew what electricity was actually useful for. Until Michael Faraday came along. In 1821 he made the great leap forward of finding something to do with electricity when he built the world's first electrical motor. This made him famous and made his old friend William Hyde Wollaston furious – he claimed it was his idea. What Faraday did work out on his own however was that by cranking his electrical motor by hand, it worked in reverse, producing an electrical current – the generator had been born.

All he needed now was somewhere to store all that electricity he was producing and fortunately, across in Italy, a friend of Galvani – Count Allessandro Volta – had just conveniently enough invented the battery or 'Voltaic pile' as he called it – the successor to the Leyden Jar. At the same time a French mathematician who had made his name by proving that habitual gamblers always lose in the end (something which no one seems to have realised before) developed the mathematics behind this strange electrical–magnetic force and in gratitude had the unit of current – the ampere – named after him.

Now that electricity could be generated and stored it opened the floodgates to new electrical inventions. Only one question remained. How to get this wonderful new power to the masses. This would involve a man who invented a death ray, a horrific form of execution, and an elephant called Topsy – but that is another story.

Episode Sixteen – The War of the Currents

In fact, it's this story. The War of the Currents of the late nineteenth and early twentieth centuries wasn't fought between nations or armies but between the companies of George Westinghouse Junior and Thomas Edison. In the race to provide America with electric lighting, two methods of transmission had been proposed, the 110-volt direct current model of Edison, and Westinghouse's alternating current, developed for him by perhaps the greatest electrical genius of the century, Nikolai Tesla.

Tesla was really quite clever. He spoke eight languages, demonstrated wireless communication before Marconi and invented the AC induction motor. His numerous patents contributed to the establishment of remote control, robotics, nuclear physics, computer science, X-rays and radar. He also experimented with wireless power transmission, coming up with an idea for a directed energy weapon or 'death ray' as the papers jauntily called it. He was, however, extremely eccentric, living in hotel rooms (but only if their number was divisible by three), hating round objects and refusing to touch any human hair other than his own.

He also hated Edison. Tesla had worked for the Edison Telephone Company in Paris, and in 1884 was sent to New York to meet the great man, with a letter of introduction and four cents in his pocket. He offered to overhaul the Edison company's turbines and Edison promised him $50,000 in return. But on completion of the job Edison refused to pay the fee, saying 'Tesla, you don't understand our American humour'. What with this and Edison's commitment to DC, Tesla, not surprisingly, went to work for Westinghouse.

Their AC system had the advantage of losing less power in transmission as transformers could be used to step up the voltage for distribution and then back down for public consumption, but Edison believed that this high voltage (3,000 V) transmission was dangerous. Edison's 110 V system certainly wasn't dangerous but it failed to deliver power over distances greater than about a mile without significant voltage drop. Now safety is a good ally but not nearly as powerful as convenience and economy and with them, Tesla and Westinghouse were winning the war.

Edison was not a man to sit back and lose gracefully however, so he started a campaign to persuade the public that AC was simply too dangerous to adopt. To do this he presided over a number of public electrocutions of stray cats and dogs until in 1903 the chance came to make a really big statement. That year Topsy, a circus elephant who had killed three people, was ordered to be put down. The American Society for the Prevention of Cruelty to Animals considered hanging to be too cruel and so Edison offered to electrocute the beast. The *Commercial Advertiser* of New York carried the story on Monday 5 January 1903.

Topsy, the ill-tempered Coney Island elephant, was put to death in Luna Park, Coney Island, yesterday afternoon. The execution was witnessed by 1,500 or more curious persons, who went down to the island to see the end of the huge beast, to whom they had fed peanuts and cakes in summers that are gone. In order to make Topsy's execution quick and sure 460 grams of cyanide of potassium were fed to her in carrots. Then a hawser was put around her neck and one end attached to a donkey engine and the other to a post. Next wooden sandals lined with copper were attached to her feet. These electrodes were connected by copper wire with the Edison electric light plant and a current of 6,600 volts was sent through her body. The big beast died without a trumpet or a groan.

Edison filmed the event and even went as far as trying to persuade people to refer to electrocution as being 'Westinghoused' but it was all too late. Due to the physical limitations of the DC system the War of the Currents was in fact already lost and Tesla and Westinghouse were triumphant.

But one macabre side effect of the War means it is still claiming victims to this day. In his desire to discredit the AC system Edison employed Harold P. Brown to develop the electric chair – an AC-based execution system for criminals which he hoped would satisfy the American public's desire for a more humane form of execution but also discredit Westinghouse to boot.

It was first used on William Kemmler in 1890 utilising one of Westinghouse's own generators which Edison had bought by subterfuge, supposedly for use in a university. The execution was botched and the initial 1,000 volt charge failed to kill Kemmler. He was then shocked again, this time with 2,000 volts which caused blood vessels in his arms to burst and his body to catch fire. The whole event took eight minutes and was described by one witness as 'far worse than hanging'. Westinghouse is reported to have commented 'They would have done better using an axe.' Despite this, both the electric chair and the AC system remained in use, much to Edison's annoyance.

Episode Seventeen – The War of the Knobs and Points

Benjamin Franklin was not all that averse to getting into wars – he had, after all, been a signatory on the American Declaration of Independence – but he could hardly have expected his invention of the lightning conductor to spark its own conflict. And yet that is exactly what it did.

During his researches into lightning, Franklin had notes that conductors with sharp points at the end appeared to discharge electricity over a greater distance and more quietly than those with blunt ends. From this he had reasonably deducted that the most effective way of protecting a building from a lightning strike would be to attach a sharp iron point to the highest part of the building and attach it to a wire that conducted the charge to earth. He of course put it more elegantly than this, saying:

the electrical fire would, I think, be drawn out of a cloud silently, before it could come near enough to strike and thereby secure us from that most sudden and terrible mischief.

The idea caught on quickly in the nascent USA, the Philadelphia State House soon sporting its own gilded iron lightning conductor, but was less well received abroad. In France the electrical experimenter Abbé Nollet, who had demonstrated electrical transmission by making a line of minks hold hands and then discharging a Leyden Jar into one of them, wondered if conductors might actually attract lightning.

The English disagreed, as they did with almost anything the French were wont to say, and after St Bride's Church in Fleet Street, London, was severely damaged by a lightning strike to its spire, St Paul's Cathedral was quickly fitted with one of Franklin's 'points'.

But it was not only church builders who were worried by lightning. No one feared electrical storms more than the builders and users of gunpowder stores – or indeed anyone living within a few miles of one. It was this that brought Franklin's point to the attention of the learned gentlemen of the Royal Society who set up a commission into the subject which suggested the installation of pointed conductors on the roof of the gunpowder magazine of the Board of Ordnance House at Purfleet.

This was an excellent idea but had a rather tragic outcome. The conductor was duly fitted but not long after, in 1777, the magazine was hit by lightning. Now I should add that the magazine did not explode, indeed only a few bricks were dislodged so it might at first glance seem that the conductor had already proved its worth. This, however, was not what committee member Benjamin Wilson thought.

Wilson had disagreed with the committee, not about the efficacy of lightning conductors, but about their shape. He partly agreed with Abbé Nollet that pointed conductors might actually attract lightning or as he put it might cause 'the very mischief we mean to prevent'. His solution was to use blunt rods, preferably with 'knobs' on the end. To give the Royal Society its due, it had then conducted experiments with blunt conductors and then begged to disagree, much to Wilson's fury.

Such academic arguments are the bread and butter of research so all this might seem to be something of a storm in a teacup. But against the backdrop of the American Revolutionary War Wilson found a new ally in his misguided quest – patriotism. Franklin, as we have noted, was a signatory to the Declaration of Independence and as such, in Wilson's eyes at least, a traitor whose scientific opinion on anything must hence be suspect. Using this logic he masterfully managed to conflate the idea of supporting pointed lightning conductors with supporting the American revolutionaries. In this mad world, just suggesting using a pointed conductor was tantamount to siding with the rebels. As Wilson put it, Britain was duty-bound to 'discard the invention of an enemy'. Knobs were patriotic, points were treacherous.

Sadly this madness came to the attention of a man for whom a form of madness was a way of life – King George III. He agreed with Wilson and ordered all 'points' be removed from government buildings and replaced with 'knobs', although, as with many of George's orders, it's unclear if the order was ever carried out. To show his further displeasure with the clearly rebellious Royal Society he summoned its president, Sir John Pringle, to explain why the learned gentlemen had so outrageously sided with the enemy and to demand that they recant. Pringle nobly, but suicidally, declared:

My duty, sire, as well as my inclination, would always induce me to execute your Majesty's wishes to the utmost of my power, but I cannot reverse the laws and operations of nature.

George, who saw no reason why the laws of nature shouldn't bend to his will, promptly had him fired both from his job at the Royal Society and as a royal physician.

Franklin meanwhile was taking the 'war of the knobs and points' with all the seriousness it deserved, remarking that he rather hoped King George would dispense with lightning conductors altogether as it was only thanks to them that he was safe from a vengeful God who would otherwise have struck him down with a well-aimed bolt for attacking his American subjects.

In the end no divine intervention was needed and both the Revolutionary War and the correct shape for lightning conductors were concluded in Franklin's favour.

A friend of Franklin later summed up the view in this splendidly treasonable verse:

While you, great George, for knowledge hunt,
And sharp conductors change for blunt,
The nation's out of joint;
Franklin a wiser course pursues,
And all your thunder useless views
By keeping to the point.

Episode Eighteen – The Beer of Revenge

I'm fond of beer and, in small quantities, it's fond of me, but I have found myself wondering why so much of the beer we drink today is lager? That is all thanks to a nineteenth-century war and a bit of vengeful chemical engineering.

Louis Pasteur is undoubtedly one of the greatest names in science – after all, how many of us can say our name is on every pint of milk? But this doesn't mean he was necessarily a very nice person. What particularly got Pasteur hot under the collar was Prussia and all things German, in large part due to the Franco-Prussian War whose outbreak in 1870 brought the building work on his new laboratory to a grinding halt. Of course it's perfectly acceptable to take against people who invade your country and stop what would prove to be life-saving work on germ theory, particularly when the ensuing war sees your son enlist and fall dangerously ill with typhoid, but Pasteur took his hatred to extremes.

His unquenchable abhorrence of all things Prussian took two visible forms. Firstly he insisted that every paper he published would contain the statement 'Hatred towards Prussia! Revenge! Revenge!', which must have proved difficult for peer reviewers, but had little real impact. But the second form changed beer drinking as we know it.

Until the 1860s beers were generally dark liquids in which yeast fed on sugars from malted barley to produce alcohol and carbon dioxide, the bubbles of which made the yeast float on the top – what is known rather unimaginatively as 'top fermentation'. But in the 1860s German brewers had engineered a small miracle. They cultivated yeasts which acted very slowly at very low temperatures and sank to the bottom. You can guess what this process was called. These yeasts produced a lighter, straw-coloured beer that kept exceptionally well and was dubbed 'lager', German for 'storage'. The result was a massive expansion in German beer production and its export across Europe. This was where Pasteur decided to strike. Taking his mind momentarily off protecting the world from deadly disease he turned his genius to engineering the finest beer in the world, a beer that would be so good it would destroy the German brewing industry and which he would dub 'The Beer of Revenge'.

Pasteur's plan was simple – well, simple for a man as brilliant and driven as Pasteur. He began isolating yeast strains which behaved like the German bottom-fermenting varieties but which acted faster and were more temperature tolerant so, in an age before refrigeration, could be used in climates warmer that Germany's. He also turned his mind to preventing spoilage in his 'superbeer', using the skills developed in isolating anthrax to isolate the spoiling organisms that contaminated beer and then designing and patenting machinery and systems to keep them out of the brewing process. In all this work he was assisted by another great man with a very arduous job. Pierre-Augustin Bertin was the Head of Scientific Studies at the École Normale and had one key advantage over Pasteur in this work – he actually liked the taste of beer. So his role was to taste every beer that Pasteur brewed up to ensure that not only would it keep but that it was actually pleasant to drink.

After much experimenting by Pasteur and an inordinate amount of drinking by Bertin, the type and process for making the Beer of Revenge were perfected. Pasteur now had his weapon and he immediately put it into action. Arranging a tour of European breweries which notably excluded any in German territories, he began freely sharing his secrets, notably with the Carlsberg Brewery in Denmark and Whitbread in England, although anyone could use his techniques provided they weren't German. The result was spectacular for, although brewing was an age-old practice, very few brewers knew exactly what happened to turn their ingredients into beer. Pasteur did know and he exhorted brewers to buy a microscope so they could examine their yeast and identify the thread-like fungi that often infected and spoiled it as well as recommending techniques for growing pure yeast strains and preventing their infection. He also wrote a book on the subject, which became something of a brewers' bible but with the strict caveat that it must never be translated into German, which it wasn't.

So did Pasteur get his revenge? Well it would be wrong to give all the credit for the brewing revolution to Pasteur but his Beer of Revenge did have a profound effect on the German brewing industry as other nations now had the techniques – and better ones at that – for making their own beer. These new, easy to produce and long-lived lagers would go on to take over the brewing world, hence their popularity in pubs to this day. But as is so often the case with history, there was one small, unintended side effect. With many German breweries now idle, their owners had to turn their machinery to other uses and found the perfect product in acetone. The acetone was needed for the production of cordite for the ongoing build-up of German armaments that would later be unleashed on Pasteur's homeland at the outbreak of the First World War.

Episode Nineteen – No More Heroes
(or why timing in engineering is everything)

If there's a single iconic invention in the history of engineering and technology it has to be the steam engine. Its story is oft told and every British schoolchild knows it was invented by Thomas Newcomen. Or was it? Newcomen was an inventor in the days when, unless you were fabulously wealthy with a lot of time on your hands, being a 'scientist' wasn't really rated. Lacking that fabulous wealth and coming instead from the rather more modest Devonport family he was what was known as a 'schemer'. The fabulously wealthy liked to refer to these people as 'blacksmiths' to remind them they were simply workers although he preferred the slightly posher term 'ironmonger', in the process displaying an early case of 'status anxiety'. Newcomen wanted to study the problem of pumping water from mines and it is in solving this problem that he is credited with inventing the steam engine. Except when he got to Cornwall, where deep tin diggings had just this problem, what did he find there? A steam engine.

This device was the work of one Thomas Savery, who had rather pre-empted Newcomen and had managed to get his work published. Being of a somewhat grander class that Newcomen, Savery was no schemer but instead could call himself a 'military engineer', as indeed he did in his book *An Engine to Raise Water by Fire*. Not that Savery had it all his own way. There are always people posher than yourself and when he suggested to the Admiralty a novel idea he'd had for propelling a ship using paddle wheels

– an idea we now know to be rather ahead of its time – they replied rather sniffily that they couldn't quite see why they should:

... have interloping people, that have no concern with us, to pretend to contrive or invent things for us?

Savery to his great credit, built his ship anyway and cruised it majestically up and down the Thames. Sadly for him, no one important was watching, so that was that. And that is how he came to turn his attention to pumping out mines and hence brought him to the invention Newcomen was now looking at.

So Savery invented the steam engine. Well ... a century before Savery, Edward Somerset, 2nd Marquis of Worcester, had been pondering the knotty problem of perpetual motion and suggested a 'semi-omnipotent and water commanding engine' in his own book. Jean Theophile Desaguliers (who despite his name was English and who also has a claim on inventing the steam engine) claimed that Savery read Worcester's book, and then bought up all the copies available and burnt them so that he could claim to have invented the engine himself. Even if we allow that this might just be sour grapes on the part of Desaguliers even Savery couldn't deny that he'd 'borrowed' the idea for a steam boiler from Frenchman Denis Papin who described his 'Digester' or 'pressure cooker' in 1679.

But if there were precedents to Savery's work, he can at least be said to be the first person to actually have tried to build one of these engines. And that was perhaps part of the problem. As the first real steam engineer, every unsolved problem with the technology fell in his lap. Despite getting a patent from King William III, who was rather impressed when shown a model at Hampton Court in 1689, the real thing proved hard to build. His engines were weak and required huge quantities of fuel for very little power gain. They were also subject to dangerous modification. Savery had, thoughtfully, included a pressure valve in his design (invented by Jean Theophile Desaguliers) but mine owners keen to improve the work rate, as mine owners always are, had a habit of putting weights on the valve to increase the pressure and hence power of the engine. I don't need to tell you the usual result. So, when all was said and done, most owners considered that the traditional donkeys and buckets used to pump out mines for centuries were still, sadly, more efficient. It probably didn't help that he chose to call his machine a 'fire engine' either.

So what was Newcomen's bright idea as he sat looking at Savery's hissing monster of a machine? His revelation was that Savery had got the whole thing upside down. Savery used steam pressure to push a plunger up a piston which required high steam pressure in boilers which tended to lead to explosions as welding was still more of an art than a science. Newcomen wondered if the opposite wouldn't work better. Newcomen decided to fill a piston with steam at low pressure and then suddenly cooled it, by drenching the system in cold water. This made the steam condense out, creating a partial vacuum in the piston tube which led to atmospheric pressure outside pushing the piston back down. Atmospheric pressure is, of course, enormous so this 'atmospheric engine' proved much more powerful and much safer.

And so we can imagine how the other schemers and engineers of the day beat a path to his door to congratulate him. Well we can *imagine*. In truth they simply didn't. Desaguliers was forced to admit that Newcomen and his partner John Calley had found a good solution but added:

not being either philosophers to understand the reason, or mathematicians enough to calculate the powers and proportions of the parts, they very luckily, by accident, found what they sought for.

The great scientist Robert Hooke, who should have known something about having a career blighted by jealous contemporaries, was no more encouraging. When Newcomen explained the project to him he told him in no uncertain terms to simply stop. It would perhaps be worth mentioning at this point that Hooke's close friend, the Dutch astronomer Christian Huygens, had his own design for an engine in the offing, although his, rather alarmingly, used gunpowder instead of steam to drive it.

By now you're probably wondering who then actually did invent the steam engine, and the answer is simple – none of the above – for the steam engine had been invented over 1,600 years before. The only problem was no one wanted it. The first steam-powered device wasn't a simple steam engine, it was a fully formed jet turbine, invented sometime in the first century AD by Hero of Alexandria. It consisted of a boiler leading to a freely rotating copper sphere with two angled outlets. As the pressure built up in the sphere, steam shot out of the angled nozzles, making the whole thing spin round like a Catherine wheel. At the time it was the fastest moving man-made object in the world. So impressed were the Romans who then ruled Egypt that they totally ignored the device. Had they taken it seriously the industrial revolution might have started two thousand, rather than three hundred years ago.

Hero's problem, like so many other inventors of the steam engine, was not a matter of what he did but when he did it. Romans had slaves to do the donkey work and the last thing they wanted was a machine that made slaves redundant and gave them time to, say, revolt. So Hero's aeliopile remained just a toy to amuse Roman dinner guests. He had simply invented it too early, leaving Newcomen to take the laurels a millennia and a half later – which just goes to show that when it comes to engineering, timing is everything.

V

In which we meet a nice arms dealer, build a ship out of ice and get Christmas off with a bang.

Episode Twenty – The Shoreham Mystery Towers

Everybody likes a mystery, provided they're not paying for it, and few in engineering terms have been as strange or costly as the story that began in June 1918 with the arrival of a detachment of Royal Engineers at Southward Green in Sussex.

The work party that summer was all sworn to the utmost secrecy as it set about building a camp to house the staff for project 'M-N'. Soon after, the locals in Shoreham couldn't help but notice that a huge construction project had begun in the harbour as 3,000 men laboured, mainly at night, on two colossal but mysterious concrete and steel towers.

If the interest of the locals was piqued they knew better than to ask – it was wartime after all and the work was obviously secret, hence the night-time working. But they could hardly ignore the towers. Each one stood on a hollow, 80-foot thick concrete base and the 40-foot wide, 90-foot high, 1,000-ton steel column that emerged from this was surrounded by a lattice of steel bars. By the autumn of 1918 they had a nickname – 'the Shoreham Mystery Towers' – and newspapers from as far afield as New York were speculating as to their purpose. They were now so large that you could have seen them from as far away as Beachy Head.

There was also a lot of speculation back at the Admiralty in London – not about their purpose – but about whether they were worth the effort. The towers were costing well over £1 million each (or, using average earnings for comparison, £172 million in modern terms) and eight (some sources say twelve and one sixteen) of these towers were on the planning boards. No one was sure if the rest could even be afforded.

So what were they for? Project M-N had been initiated by Sir Alexander Gibb who was then engineer-in-chief to the Admiralty. Gibb had engineering in the blood, indeed the profession was so ingrained in him he might as well as have oil for blood. His father was the founder of the engineering group that would become Easton, Gibb and Son, his grandfather had been a pupil of Telford, his great-grandfather was a founder member of the Institution of Civil Engineers and his great-great-grandfather was a contemporary of Brindley and Smeaton. He had worked as resident engineer on the Whitechapel extension to the Metropolitan line before the First World War and overseen the construction of Rosyth naval base before being appointed chief engineer ports construction, to the British armies in France in 1916 with special responsibility for rebuilding the railheads and ports that might be destroyed by retreating Germans. Then in 1918 he had come to the Admiralty, where his fabulously expensive towers were causing something of a headache.

Gibbs' plan was typically bold. With the brilliant but untrained Major John Reith (later Lord Reith, first Director General of the BBC) as his assistant, he intended to nullify the menace of German U-boats in the simplest way possible. He planned a chain of huge towers across the entire navigable width of the English Channel from Dungeness to Cap Gris Nez, each linked to the next by steel anti-submarine nets, effectively closing off the whole of the world's busiest seaway to U-boats. Each tower would further have a steel superstructure containing gun emplacements and with room for 100 troops to man each lonely outpost. They would be equipped with submarine detection equipment run off their own generator. Any U-boat captain brave enough to try to enter the Channel would get snagged in the nets and be either trapped under water there or surface and be shot to pieces by the tower crews.

It was certainly one of the most ambitious engineering plans of the war but it was a plan too late, much to the relief of the Admiralty accountants who were looking at a £2 billion bill (in modern terms) to complete the project – enough to sink a few battleships, or perhaps float a bank. On 11 November with Tower 1 just nearing completion, but still unfinished, the war ended and the behemoths became redundant overnight. Tower 2 was eventually broken up for scrap in 1924, a task which took nine months – longer than it had taken to build – but Tower 1 is, rather splendidly, still visible today as a new life was found for it.

In early September 1920 the honeycomb-structured concrete base was pumped free of water and the 30,000-ton structure rose free from the Shoreham seabed. On Sunday 12th the tower was then towed out of Shoreham harbour by Admiralty tugs, only just clearing the harbour walls. After a journey of 41 miles, by the Nab Rock, off Bembridge on the Isle of Wight, the tower was tethered and the seacocks in the hollow base opened. Slowly the tower began to sink, coming to rest on a sand bank at a jaunty angle of three degrees from vertical. The new Nab Tower lighthouse was now in place and the four crewmen of the old Nab Rock lightship could be transferred to their new, hi-tech, multimillion pound home. And not surprisingly for a 30,000-ton structure, that is exactly where it remains to this day. Now solar powered and unmanned, Tower 1 still serves as the Nab light and as a reminder of what was once the Admiralty's greatest secret.

Episode Twenty-One – Bang Goes Another Good Idea

Opinions of arms manufacturers tend to fall into one of two camps. Those wielding their products, provided they work as advertised, tend to be all in favour of them whilst those who find the products pointing at them are decidedly less enthusiastic. One thing that both sides in this unfortunate stand-off would agree on however, is that armament manufacturers are remarkably innovative and, having both the finance and the inclination, they have, over the years, made the odd breakthrough that has been of great benefit to humanity – even if this was accidentally.

All of which brings me to Sir William Armstrong. Armstrong is probably best known today as founder of one of Britain's most successful armaments companies and the engineer responsible for the invention of the Armstrong gun. Indeed his creation of the first large, rifled artillery pieces, to replace cast-iron cannon, would ensure his fortune, greatly bolster the prospects of the British Empire and result in him becoming the first engineer ever to be ennobled. A grateful government even took out patents on his invention in his name (Armstrong wasn't a great fan of patents) and then prevented their publication so as to ensure foreign powers wouldn't get their hands on his great ideas.

But there was another side to William Armstrong, that has been rather lost in the fog of war. Armstrong had not set out to be an armament manufacturer. In fact what really interested him was water and it had done ever since, on a fishing trip in his youth, he had watched an overshoot waterwheel and marvelled at its inefficiency. As he put it:

I was lounging idly about, watching an old water-mill, when it occurred to me what a small part of the power of the water was used in driving the wheel, and then I thought how great would be the force of even a small quantity of water if its energy were only concentrated in one column.

Inspired to create better devices for using water power he had invented a water-powered rotary engine and, undeterred when no one took an interest in this, went on to found a business making hydraulic cranes. This then was what Armstrong thought his business career would be all about – renewable energy.

If would be fair to say that the mid-nineteenth century was not an era where renewable energy sources were viewed as particularly important. Unless you were William Armstrong that is. Armstrong was one of the first industrialists to realise that our dependency on coal was using up a finite resource. In his Presidential address to the British Association for the Advancement of Science in 1863 he rather presciently predicted that Britain would cease to be a coal-producing power within 200 years, a theme he continued to develop for many years after. As he warned the association, coal was 'used wastefully and extravagantly in all its applications'.

What Armstrong was interested in was power that would last forever. He suggested that solar energy would one day replace the steam engine and calculated that solar energy falling on an acre of the tropics could usefully produce 4,000 horse power for nine hours a day – although he didn't have a method for extracting this energy. What he could harness however was water power and his monument to this would be his own house at Cragside in the Debdon Valley. It was a site that he had visited many years before and, taking his first holiday for many years there he found the air healthful and, more importantly, the waters of the Debdon Burn ripe for tapping.

What began as a small hunting lodge soon developed into a magnificent industrialist's mansion, designed by R. Norman Shaw. But it was what was inside that had government ministers and royalty itching for an invitation. Armstrong had dammed the Debdon Burn high in the hills, creating a head of water (later greatly added to) that he put to use powering labour-saving devices in his house. There was a hydraulic spit turner, a servants' lift and, remarkably, a dish washer. Leading on from his early interest in electrostatics, Armstrong also harnessed water power to generate electricity, making Cragside the first house on earth to benefit from hydroelectric power. It was this that really made the house a marvel, with its electric dinner gong, fire alarm, internal telephone system and electric arc lighting. For Christmas 1880 the whole of the lighting system was then upgraded with the world's first full installation of his friend Joseph Swan's incandescent light-bulb system. One visitor to the house summed up the general view when he said that Cragside was 'the palace of a modern magician'.

Undoubtedly the wonders of Cragside were partly there to impress the visiting kings and princes who were interested in buying guns and warships from Armstrong, but his interest in renewable energy was genuine. At the annual meeting of the British Association in 1883, Armstrong spoke again as president and took the opportunity to hypothesise how thermoelectric energy might one day replace steam and how solar heat and hydroelectricity would one day come to be vital, also prefiguring the construction of a National Grid.

Armstrong died childless in 1900 and for much of the twentieth century his house and his ideas fell into disuse – although his guns did not. With the gifting of Cragside to the National Trust in 1977 the house has at least been brought back to its former glory. In an age when we may need all the engineering genius we can find to solve our energy problems, let's hope the 'other' side to William Armstrong will again come to the fore.

Episode Twenty-Two – The Ice Man

Snow is something of a mixed blessing. Headline writers wallow in it, people with places to go hate it, but it's great fun for those of us with children or childish interests, of which I have both. And so, at the first sign of a flurry, my thoughts turn quite naturally to making things from it – snow balls, snowmen and snow angels. But not aircraft carriers. That's the difference between me and tragic genius Geoffrey Pyke.

Pyke was one of Britain's greatest eccentric engineers, a man who cut his inventive teeth as a First World War POW making a methodological study of how to escape his prison – something then thought largely impossible but which he triumphantly did. He arrived back in Britain a hero and promptly arranged for the details for his escape plan to be sent to his fellow prisoners in false-bottomed boxes. They assiduously ignored these and stayed put, proving that escaping is indeed impossible for those who don't want to.

At the end of the war Pyke turned his unusual, analytical mind to the problem of making money, and he came up with a typically risky but successful solution. He would mathematically analyse movements in the commodities markets, mainly metals, and through a series of brokers (so no one knew how large his placements were) he'd hedge against their movements. In other words he'd run a hedge fund, which he did very successfully right up to the moment it went bankrupt, which is a phrase that seems to have a certain resonance again at the moment.

But Pyke wasn't a grasping plutocrat: far from it. He'd invested all the money he'd made in his own school in Cambridge which was designed to encourage junior school children to learn by themselves – encouraged by adults, not lectured by them. Sadly this went bust along with the business. If Pyke was down he was not out however. At the outbreak of the Spanish Civil War he put his inventive mind to work combating the rise of fascism by arranging the export of dried sphagnum moss to make cheap wound dressings.

A larger war was by this time in the offing and just before it broke out Pyke had another brilliant idea. He would arrange for a team of opinion pollsters to go to Germany disguised as a golfing tour. In their matches they would casually ask locals what they thought about the idea of war. These results would be collated and then sent to Hitler to persuade him that his people had no stomach for it. It was a madly brilliant idea but sadly the war broke out before the results could be sent to the Führer.

Pyke had now hit a purple patch in his unorthodox thinking and his strangely wonderful ideas began to flow more freely. He posited the forerunner of the skidoo, which would allow small numbers of commandos to travel quickly around occupied Norway, tying up thousands of slow-moving Wermacht troops trudging through the snow. All this thought of snow and ice then gave Pyke an even greater idea. How do we protect the Atlantic convoys? Aircraft. But how do we get them to the mid-Atlantic? Aircraft carriers. But they needed to be very large and hence they were obvious, sluggish targets. And even the largest were nowhere near large enough to carry most fighting planes. The answer? Build them out of ice. So project Habbakuk (the spelling comes from a Civil Service typing mistake) was born. Pyke's ships would be huge – twice the size of the largest liners – but made of ice. Torpedoes would find them almost impossible to destroy and any holes could be repaired with water.

Having taken the idea to the refugee Austrian chemist Max Perutz at the Cavendish Laboratory in Cambridge, he suggested that ice mixed with wood powder would be even stronger and better resist deformation (creep) and melting. This new material, christened Pykrete, as it was as strong as concrete, soon got the blessing of Churchill, allegedly after Mountbatten threw a lump of the wonder material in his bath. So a small test rig was built on Patricia Lake, Alberta, in Canada, which proved to be a great success. Pyke returned from Canada elated but already events were again overtaking him. The U-boat threat to the Atlantic convoys – the ideal location for an iceship – had faded and the action was turning to warmer and hence less suitable waters. Also the improved range of aircraft made vast mid-oceanic landing strips less necessary. So Pyke's greatest idea was shelved. He did suggest smaller versions, perhaps as temporary harbours, but the Mulberry scheme was employed instead.

After the war Pyke went back to trying to improve the common lot. He was a committed campaigner against the death penalty, and personally helped to humiliate the British government into supporting the newly-formed United Nations International Children's Emergency Fund (UNICEF), which it had done its best to ignore. At his death he was working on modelling how the country would pay for its nascent National Health Service but a pervasive sense of gloom seems to have come over him. His ideas were widely ridiculed in the press – he was the 'mad boffin' whose ludicrous ideas only demonstrated their inventor's distance from reality. On the evening of 21 February 1948 he took a fatal overdose of barbiturates and died in his sleep.

Episode Twenty-Three – Spring Time

Whilst we're on the subject of snow it would be criminal not to tell the story behind a very Christmassy piece of engineering and one which may well have sparked the interest of many thousands of children in a future career in engineering itself.

Richard James had not intended to become a toy maker. Indeed in 1943 you would have found him at work in a machine about as far from a toy as you might imagine – a battleship then in Philadelphia's Cramp shipyard. Down in the engine room James had been given the job of finding a way to mount very sensitive horse-power meters in a room that, under full steam, would shake so much your fillings would fall out. To do this he was experimenting with looms of torsion springs which he hoped would damp the vibration. During these experiments he accidentally knocked one of the springs off the table and noted something very odd. Instead of just clattering to the ground, the spring seemed to 'walk' off the edge of the table and onto the floor.

Like all good engineers, James saw an interesting opportunity when it walked off a table in front of him. Going home to his work room (where he had already installed probably the world's first on-demand iced coke system, powered by a compressor that pumped coke straight from his basement into his fridge) he set about experimenting with various types of spring, using different steels and coiling them to different tensions. When he was happy with the result he showed the 'walking spring' to his neighbours' children and the reception was so positive that even his wife Betty, who was rather used to Richard's 'great ideas', agreed that it was worth investing in.

Not being a family of great means, Richard and Betty took out a $500 loan to found 'James Industries', promptly spending nearly all the money on having a local machine shop produce 400 of the springs. The rest was spent at the printers producing a printed instruction sheet which Betty hand-wrapped around each spring as its only packaging.

All that remained was to find a name for the toy and some buyers. The first problem was solved by Betty who searched the dictionary for a word that summed up the sinuous, sleek movement of the spring, finally stumbling on the word 'Slinky'. The second problem might have proved a bit trickier. Fortunately Richard managed to persuade the owners of Gimbel's department store to let him take a demonstration pitch in their toy department and in December 1945 he set up his stall there, consisting of a wooden board propped up at an angle and 400 neatly wrapped springs.

It was a gamble and Betty was keen to reduce the odds so she and a friend waited outside with a dollar bill in hand intending to rush in when Richard was set up and make the Slinky look more popular than it was. In fact she needn't have bothered. By the time she reached the Slinky stand Richard was already surrounded by eager customers trying to force their money on him. In just 90 minutes those first 400 Slinkys sold out and one of the iconic toys of the twentieth century was born.

The only problem that now faced James Industries was that they had no industry. Betty set to work on the packaging, producing a simple box to hold the toy whilst Richard used his engineering knowhow to design and build a machine that could coil the 80 feet of wire in a Slinky into the requisite 98 coils in just under eleven seconds. By 1950 the toy was so successful he'd had to build another five machines, all of which are still in operation today. In that time the only alterations to the product have been the introduction of crimped ends (for safety reasons) and the transition from Swedish blue steel to a cheaper American steel.

Fame and fortune certainly followed although in 1960 Richard suddenly left the business, Betty and his six children to join what the family called a 'religious cult' in Bolivia. With him went a lot of the money from the business, much of which he gave to his new Bolivian friends. Perhaps appropriately for a spring manufacturer however, Betty sprang back, took over the management and paid off the company debts. Moving the toy business back to her appropriately-named home town of Holidaysburg, she negotiated her way out of a supply crisis caused by striking steel workers, and introduced new innovations such as the goggle-eyed Slinky glasses, the Slinky dog (made yet more famous by the Toy Story movies) and the plastic Slinky. In 1998, aged 80, she finally agreed to sell the company so she could spend more time with her family. There is a Slinky in the Smithsonian, a Slinky has been in space, and the toy has even been honoured with its own US Postal Service stamp. But most importantly of all, over 300 million of these simple engineering components have been sold to children across the world, contributing to many hundreds of thousands of very happy Christmases for everyone except the grown-up given the job of untangling the thing ...

Episode Twenty-Four – Father Christmas

Many a Slinky has found its way under a Christmas tree but Christmas trees have always presented something of an engineering problem. Their introduction into British living rooms is often attributed to Prince Albert but in fact the royal family had been using them since George III's reign when his Queen, Charlotte of Mecklenburg-Strelitz, introduced this German tradition to brighten up the festive period and probably also to take her mind off the fact that her husband had taken to chatting to trees in Windsor Great Park.

But bringing a tree into your house is tricky. In the first instance, millions of years of evolution have accustomed pine trees to living outdoors in an environment not really very much like a suburban living room. Secondly, from the earliest days it was thought advisable to put a few lights on the branches to jolly up what might otherwise look like a small and gloomy forest glowering in the corner. The answer to the first problem was simple for Charlotte of Mecklenburg-Strelitz who had plenty of spare trees in the garden so didn't really mind if one died, and who also had 'people' to clear up the pine needles. The second problem was more intractable however. The German tradition was to put candles on the tree, but the combination of naked flames, bone-dry branches, resinous needles, paper decorations and, sometimes, the addition of fiddling little fingers, sometimes left more than chestnuts roasting on the open fire.

It's at this point that we could do with an engineering hero and sure enough, just as firemen are beginning to think they may never get a Christmas Day off again, Edward Hibberd Johnson comes on the scene. Johnson had been assistant to railroad pioneer William Jackson Palmer and it was when his boss sent him back from the railhead to run the Automatic Telegraph Company that he came across a likely lad to employ called Thomas Alva Edison. Impressed with his huge work rate and brilliant ideas, Johnson and his boss decided to set Edison up as a full-time inventor starting him off on the road that would lead to the invention of the carbon microphone, a reasonably reliable light bulb, a power generation and distribution system and the phonograph.

Not content with recognising one electrical genius however, Johnson then went on to hire Frank J. Sprague, a naval officer whom he persuaded to retire from the military for a life of invention. Sprague would go on to invent the pole which connected trolley buses to overhead power supplies and be instrumental in the development of electric elevators. In this way Johnson seeded the now famous Menlo Park laboratory with the people who would go on to make the Edison General Electric Company a household name – a household name of which he would be a director.

The rest of course is history. Edison went on to be called the 'Father of Electricity' – although I expect Michael Faraday amongst others might have had something to say about that. Frank Sprague went on to be called 'the Father of Electric Traction' (which was perhaps less disputed) but what of poor old Edward Johnson? Well, Johnson is the hero of this fairy story and he's known as the father of something too. He might not have had the genius for invention of an Edison or the mathematical intuition of a Sprague but he did manage a first that every fireman who wanted to eat his Christmas dinner in peace, and child who ever looked with wonder on a Christmas tree, would thank him for.

On 22 December 1882, just three years after Edison's first successful experiments with carbon filament lightbulbs, Edward Johnson invited some friends (and a few newspapers) to his New York home for a little demonstration. A reporter from the *Detroit Post* noted:

Last evening I walked over beyond Fifth Avenue and called at the residence of Edward H. Johnson, vice-president of Edison's electric company. There, at the rear of the beautiful parlors, was a large Christmas tree presenting a most picturesque and uncanny aspect. It was brilliantly lighted with many colored globes about as large as an English walnut and was turning some six times a minute on a little pine box. There were eighty lights in all encased in these dainty glass eggs, and about equally divided between white, red and blue. As the tree turned, the colors alternated, all the lamps going out and being relit at every revolution. The result was a continuous twinkling of dancing colors, red, white, blue, white, red, blue – all evening.

I need not tell you that the scintillating evergreen was a pretty sight – one can hardly imagine anything prettier. The ceiling was crossed obliquely with two wires on which hung 28 more of the tiny lights; and all the lights and the fantastic tree itself with its starry fruit were kept going by the slight electric current brought from the main office on a filmy wire. The tree was kept revolving by a little hidden crank below the floor which was turned by electricity. It was a superb exhibition.

Of course Johnson was just making a start. Initially everyone in New York thought it was simply a publicity stunt and it took President Grover

Cleveland to light the White House tree with electric lights before the idea really took off. Even then not all of Johnson's ideas lasted. Having the whole tree revolving every ten seconds would today be considered a little alarming in most instances and as everyone who has ever bought a set of 'chasing' lights will know, even the strongest of us can have too much 'twinkling' of a evening. Other fashions have also come and gone – fibre-optic branches, reflection-lit aluminium trees (in 1950s America) and now low-energy diodes – but Johnson's basic idea has remained: to turn one corner of an ordinary room into a little bit of Christmas magic.

In this way Edward Johnson has gained the title 'the Father of Electric Christmas Tree Lights'. It is perhaps not as grand a title as Edison's or even Sprague's but one which will resonate considerably more with every ten-year-old who still appreciates the Christmas tree sparkling in the corner but gives not a jot for the memory of the phonograph or the trolley-bus.

Episode Twenty-Five – Snap, Crackle and Pop

Christmas just wouldn't be Christmas without the groaning, sighs of disappointment and faint smell of gunpowder in the air, all produced by that most British of Christmas inventions – the cracker.

The traditional cracker with its chemical 'snap', cringe-worthy 'joke', flimsy paper hat and often rather shoddy 'novelty' was not the invention of an engineer, sad to say, but rather a young confectioner. Yet its bringing together of different technologies created a marketing sensation that we might all learn from.

Tom Smith started out in 1830 in a bakery which also sold a sideline of confectionery and cake decorations. This was an era when the sweet business was not overly sophisticated and the stuff Smith was making and selling was little more than bags of gum pastilles and fondants, sold unwrapped and undecorated. It was a business ripe for development and, in 1840, Tom, having set up his own shop in Clerkenwell, took the family on holiday to the sweetie capital of the world, Paris, to search for inspiration. Here the family stumbled upon the bombonière, a sugared almond like any other but wrapped in a twist of colourful waxed paper, turning it from a simple sweet into a small present.

Tom Smith loved the idea and back in London set about making his own bonbons, which proved a great success that Christmas. After the holiday season however, demand suddenly dropped off and Tom was forced to fall back on his cakes, as it were. Clearly his customers saw his tissue-wrapped bonbons as a Christmas treat and he desperately needed to widen their appeal. So the first development in the birth of the cracker came along

– the motto. Tom cannot however be blamed for the appalling jokes in modern crackers. His idea was altogether more saucy – love mottoes. In each wrapper he placed a piece of paper with a provocative line such as:

The sweet crimson rose with its beautiful hue is not half so deep as my passion for you.

This seemed to do the trick and soon regular orders were coming in for what were, in effect, tasty chat-up lines. It was then but a short step to add a trinket or keepsake to the package to go with the protestation of love, although this of course involved a re-engineering of the whole package. Now the sweet, trinket and motto were placed in a paper tube, which was then wrapped in a twist of paper to look like a large bonbon.

There are only so many people in need of saucy sweeties however and Smith noted that he was still selling most of his wares at Christmas. It was hence one autumn whilst sitting in front of the fire, according to Tom Smith lore, that the great man then came up with his most unusual but brilliant idea. As he sat there wondering how to improve sales he kicked a log on the fire which crackled and popped. In a revelatory moment Smith realised that the logical next step in the development of his sweets was making the package burst apart with a bang to reveal its contents. So the unlikely alliance of paper, sweet, trinket and small explosive device was born. It took Smith two years to perfect the 'snap', early offerings either being so 'safe' as to be inaudible or so powerful that the cracker burst into flames. Finally in 1860 the 'Bangs of Expectation' range was launched.

These then, were the first true crackers, although they were initially known as 'Cosaques', as the 'snap' was supposedly reminiscent of the cracking sound the Cossacks made with their whips as they rode through Paris during the Franco-Prussian War. Success was instant and by 1900 Tom's factory was making 13 million crackers a year.

Nor were these filled with mass-produced bits of plastic and corny jokes. Along with the love mottoes came topical mottoes referring to current events and the arts and Shakespearean quotes. Trinkets, scoured from across the globe, included Bohemian bracelets, German scarf pins, miniature musical instruments, model landscapes and stereoscopic images presented in novelty boxes shaped as everything from a travelling trunk to one entitled 'Love in a Cottage'.

Nor were crackers then seen as purely a Christmas item. Celebratory editions came out in the 1920s to mark the discovery of the tomb of Tutankhamun (the 'Treasures of Luxor' set) and Prince Edward's world tour – 'you've met the Prince now pull the cracker!'

Bespoke crackers were also made, leading to one of the saddest tales from the Tom Smith archive. In 1927 a gentleman wrote to the company

55

enclosing a diamond ring and a ten shilling note asking that the ring be placed in a special cracker. Sadly he forgot to include his address and the ring and money still lie unclaimed, as perhaps does his love.

Since the Second World War the story of the cracker has been one of relative decline. Austerity did for the more elaborate packaging, Tom Smith's was blitzed and then bought out and recent revivals have added precious little new to the repertory. The cracker is one of those brilliantly simple, but ingenious ideas, a piece of paper engineering that has become an essential part of our favourite celebration. But no real developments have come since the days of Tom Smith himself and if the cracker is to survive perhaps it needs a new twist. Surely it can't be beyond the wit of the learned engineers to come up with an improved cracker – something to make old Tom proud.

Episode Twenty-Six – The Good Banker

I think it's fair to say that bankers aren't universally popular at the moment. In days gone by I might have said that 'their stock was rather low', but in the current economic climate their stock has all but disappeared. However, we should not immediately assume all bankers are bad – take the case of Alfred Lee Loomis (1887–1975), who might perhaps stand as a model for the current generation.

Loomis was, in his mind at least, an inventor, although he never really fitted the image of the struggling eccentric working away in his shed, eschewing meals in favour of spending the money on more raw materials. This was because he was fabulously rich. With his brother-in-law he ran the bank that raised much of the financing for the electrification of America's homes and factories and then, in a move that even made other bankers hate him, had the foresight to move out of stocks just before the great crash, returning during the Depression to buy up good companies at rock bottom prices. As I think I mentioned, this made Loomis very rich. So rich he could buy his own Americas Cup yacht, outright, at a time when even Vanderbilts and Astors clubbed together to get theirs.

So far this has precious little to do with engineering of course. A banker making a fortune and buying a yacht is, if anything, a shade predictable. But at this point Loomis and other bankers part company because he also bought something else – an engineering laboratory. Loomis had loved engineering since the First World War during which he had devised the 'Aberdeen Chronograph', a portable machine for measuring the muzzle velocity of artillery pieces. At the end of that war and with the huge increase in his fortunes, he decided to indulge this love of gadgetry by buying an empty mansion near his home and turning it into the Loomis Laboratory.

But the Loomis lab was like no other. For a start it was in Tuxedo Park, an exclusive New York enclave whose inhabitants were so rich and so posh that they still dressed for dinner, hence the American name 'tuxedo' for a dinner jacket. Secondly Loomis could afford to buy equipment that most universities could only dream of. But most importantly, Loomis was not just a tinkerer, but perhaps the last true 'gentleman scientist'. He worked with passion and skill, attracting the greatest names of the era to come and visit and even work with him. His guest book included Einstein, Bohr, Fermi and Heisenberg.

To be fair some probably came because his invitations included first-class travel, limousine transfers and the chance to arrive at the lab on his private train, but most came for the work he was doing. And that work was wide-ranging and proved to be rather important in bringing to a close a troubling little problem known as the Second World War.

Before the war Loomis worked on everything from spectrometry and chronometry to electro-encephalography and ultrasound as well as taking time out to help Ernest Lawrence raise the funding for his cyclotron particle accelerator. But with war in Europe, Loomis and his colleagues had begun to turn their attention to radio location, building one of only two microwave radar systems in the USA at that time. This got the attention of the 1940 Tizard Mission which Churchill sent to the USA to share British technological discoveries in return for American assistance. They had brought with them a cavity magnetron which had 1,000 times the power of any American machine and so Loomis invited them to Tuxedo Park to spill the beans. Shortly after this, Vannevar Bush (who, as inventor of the Memex we have already met in this book) appointed Loomis as chairman of the Microwave Committee with a remit to develop radar technology and Loomis founded a 'Radiation Lab' at MIT in Cambridge, Massachusetts – personally funded by him in its open stages whilst government financial cogs turned with their usual efficiency.

The child of the 'Rad Lab', as it became known, was the 10-cm radar – one of the key technologies of the war which helped detect U-boats, protect British and American aircraft and locate the Luftwaffe, and is often cited as the key technology for ending the war. But if you're thinking that the atom bomb was the real decider in that, then it's worth mentioning that Loomis, a friend of many Manhattan project scientists, also eased the path of that project through government (where his cousin was, rather helpfully, War Secretary).

Now it must be said that Loomis did not run the Rad Lab and so cannot be called the inventor of 10-cm radar, but his passion for the subject as a gentleman amateur, his knowledge and his willingness to use his money and influence to promote engineering, should grant him a part of the credit. And besides he did personally invent LORAN (LOng RAnge Navigation), a pulsed hyperbolic radio navigation system that allowed vessels to locate themselves using timed signals from low-frequency radio transmitters, which just goes to show that an interest in clocks combined with the ability to buy ocean-going yachts can be used to good effect. Indeed LORAN is still in use today although increasingly being superseded by GPS.

So before we sigh at the reckless behaviour of our financiers perhaps we should remember old Alfred Loomis – patron, sponsor and gentleman scientist. Let's hope our generation of bankers will be putting their bonuses to such good use.

Episode Twenty-Seven – Silent Death

Edward Harrison had not intended to join the Royal Engineers, rather he had always intended to spend a quiet life as a pharmacist, but the conjunction of the two would make him one of the most important but shamefully uncelebrated figures in modern warfare.

Harrison had shown an aptitude for chemistry from the moment he enrolled at the School of Pharmacy in 1891 and in 1894 he was elected a fellow of the Royal Chemical Society. His particular talent lay in analysing chemical concoctions and, having set up in business as an independent analytical chemist, one of his first triumphs was his British Medical Association sponsored investigation into the contents of some of the many 'proprietary medicines' then available which promised improbable cures from unlikely ingredients. His report, *Secret Remedies: what they Cost and what they Contain*, and its sequel helped to control and eventually suppress the fraudulent patent medicine market.

Harrison's career as a chemical detective looked like being cut short in 1914 with the outbreak of the First World War; indeed he was eager to give up his work in order to enlist. Sadly at 45 he was two years above the maximum age for the army and so was repeatedly turned down. In 1915 he struck lucky however – if being offered a place in the Great War can be considered 'lucky' – and, having lied about his age, was accepted as a private in the 23rd (Service) Battalion Royal Fusiliers (1st Sportsman's).

Harrison's interest and expertise in chemistry had been noted by the army and, following the first use of chlorine gas by the Germans during an attack in Belgium in April 1915, he was transferred to the Royal Engineers. They had been asked to form a Chemists' Corp with the urgent job of producing a respirator for British troops, who at the time were simply advised to put a wet handkerchief over their face in the event of a gas attack. So it was that Harrison came back from France, with a promotion to the rank of temporary second lieutenant and set to work in the supposedly much safer Millbank offices of the Royal Army Medical College.

The job in hand was both pressing and complex. The army not only required a respirator that could easily be carried by troops but it also had to be able to deal with a number of possible gasses, of which chlorine, phosgene and hydrogen cyanide were considered the most likely from a long list of over 70 possibilities. This chemical arsenal they ideally wanted countered by the impregnated flannelette hood they had already been developing.

Harrison quickly put his analytical abilities and his engineering genius to work on the problem, realising that a simple flannel hood would never be able to cope with what they were asking of it. With the aid of academic chemists at Oxford, a series of chemical countermeasures to each poisonous gas were devised and impregnated into filters. These had to be quickly tested for safety and efficiency and, against the advice of many of his more cautious colleagues, Harrison took the decision to test them on himself as he sat in sealed, gas-filled rooms.

The result of his extraordinary labours was the Harrison Tower, the prototype for nearly every gas mask that followed. It consisted of a filter box containing impregnated discs to counter each potential threat; the discs could be replaced when exhausted and changed to cope with different gases. This box was then attached to a hood with inhalation and exhalation valves.

An army service version of the Tower first appeared in February 1916 as the 'Large Box Respirator' and was immediately issued to 200,000 machine-gunners and artillerymen. A smaller version – the Small Box Respirator – more suitable for frontline infantry, was issued from August of that year and was soon part of the general issue.

Harrison could, had he been so immodest, have prided himself on the thousands of lives his respirator saved and the many more thousand terrible injuries it prevented. However, his own life was by this time tinged with tragedy, following the news of the death of his eldest son on the Somme, on 30 July – the moment of his own triumph. Harrison redoubled his efforts for the remaining years of the war but those early, desperate experiments on himself with poisonous gas had critically weakened his constitution. In October 1918 he contracted influenza and, from this, developed pneumonia which his ravaged lungs were unable to cope with. He died, aged just 49, on 4 November, a week before the Armistice was announced. Churchill, then minister of munitions, wrote a personal letter to his widow, in which he said that Harrison was about to be promoted brigadier-general in charge of all chemical warfare at the time of his death and that 'it is in large measure to him that our troops have been given effectual protection from the German poisonous gases'.

VI

In which we save things, change things and completely wipe things out. Nearly.

Episode Twenty-Eight – The Soft Machine

It's very hard to understand, and still harder to repair, a machine that you're not allowed to look at. Yet the workings of that most vital machine, our bodies, has for much of human history been deliberately hidden. In the ancient and mediaeval worlds there was frequently a religious ban on human dissection which was considered to be unnecessarily interfering in the divine work of God. Thankfully for the development of medicine there was one Roman doctor who found unique ways around this ban and discovered that the body is simply a soft machine.

Galen was born in Pergamum, in what is today Turkey, around AD 129. Galen's father, Nicon, had been a successful and affluent architect, and he had given his son a broad classical education in mathematics, logic, grammar and philosophy, but it was to medicine that young Galen turned when he was just sixteen. According to legend his father had a dream revealing his son's destiny in the subject and, whilst this would probably be enough to make most independently-minded teenagers immediately enrol in the next travelling circus, Galen took his father's advice.

The problems facing a trainee doctor in the Roman world were substantial however. Firstly one had to choose which sort of doctor to become. Physicians for the wealthy treated their patients in accordance with theoretical principles and were known as Dogmatists. Physicians for slaves relied upon trial and error, and were known as Empiricists. Between these two extremes were the Methodists, whose sole concern was the disease itself, ignoring the patient and his or her medical history. Galen however recognised that it was both necessary to develop theories from practical observation and experience and to test existing theory against observations, modifying or abandoning it if it seemed not to 'work'.

But the major problem Galen had was getting his hands on a body to experiment on. Infuriatingly, human dissection was banned in the Roman world in all but one place, so, in AD 152 he moved to Alexandria in Egypt where, thanks to the country's long history of embalming, human dissection was still permitted. His time in Alexandria gave Galen a passion for anatomy, which would be both the making of and, many centuries later, the breaking of his reputation. By opening human bodies, both living and dead, he was able to view the body as a machine and brilliantly deduce its functions. In his later career he would repeat his investigations on animals in front of astounded crowds. One slightly gothic experiment involved cutting an incision down the back of a live pig, severing groups of spinal nerves to show their function. Of course all the while the pig would squeal until he tied off the laryngeal nerve when the pig suddenly stopped. And so he proceeded, removing or tying off various vital pathways and identifying their purpose from the effect. He tied off animals' ureters to demonstrate kidney and bladder function, and closed off veins and arteries to prove that, contrary to the popular theory of the day, they carried blood and not air.

After finishing his training Galen returned to his native Pergamum where again he came up against the ban on dissection, but now he found a novel way around this, all thanks to one of the more gruesome entertainments of the Roman world.

Gladiators were the football superstars of their day, the subject of adulatory graffiti on public baths and the pin-ups of young girls everywhere. A good gladiator, although a slave, was worth a fortune to his owner and the very best doctors were hired to tend to them. For Galen this was the perfect opportunity. Gladiators often suffered horrific injuries – indeed they were almost 'self-dissecting' – and in repairing these Galen could not only make a good living but get a chance to explore inside the living soft machine. Provided it was a gladiator that made the incision in the arena and not a doctor on the operating table, no one complained if he then used these wounds – or 'windows onto anatomy' as he called them – to explore the inner workings of the body. At the same time he also pioneered the treatment of the more usual sports injuries – sprains, breaks, dislocations and concussions. From these he developed 130 of the 150 basic surgical techniques that are still in use today – everything from brain surgery to repairing compressed fractures, to the use of traction beds to straighten broken limbs. Even at the end of a gladiator's career there was one last procedure that Galen developed that would assist his patient in his new life as a free man – the removal of the tattoo that marked him as a slave.

Galen's fame spread and soon the capital, Rome, beckoned. Here he took every opportunity to put his rivals in the shade. One such occasion was provided by the arrival of a Persian merchant complaining of a loss of feeling in the ring and little fingers and half the middle finger of one hand. For some time Roman doctors had been applying unguents and creams to the fingers in the hope of stimulating them but to no effect. Galen asked him an unusual question – had he hurt his arm recently? The question must have been quite surprising. The man's problem was clearly in his fingers. What had that got to do with his arm? But Galen was right. The man said he had taken a bad fall and hurt his back. Galen diagnosed a spinal lesion and recommended bed rest and soothing compresses for the injury site. It worked and the man's fingers recovered.

Rome was good to Galen and he lived to be 87 but in the centuries that followed his death the city would also prove his undoing. Although here he could serve emperors and make a fortune he was no longer working with gladiators and he was forbidden human dissection. Instead he relied on animal dissection apparently unaware of the huge differences between the anatomy of humans and the creatures he was studying.

When, in 1543, the Flemish anatomist Andreas Vesalius, who by then had free access to human cadavers, compared what he saw in the anatomy theatre with what was written by Galen he realised that much of it was simply wrong. The inside of a dog, or even the inside of his preferred Barbary apes, was not the same as the inside of a human – and so Galen was abandoned and even his most insightful works forgotten. Galen's work, good and bad, was set aside and much of the knowledge gleaned by the first man to treat the body as a machine, was forgotten.

Episode Twenty-Nine – The Great Preserver

By the 18 February of 1807, Captain George Manby of the Cambridgeshire militia had already had a fairly eventful life. He had claimed to be a childhood friend of the greatest hero of the age – Horatio Nelson – studied at the Royal Military Academy at Woolwich, volunteered to fight in the American War of Independence (for which he was rejected on age and height grounds) and joined the militia. He had also become a published author, married a wealthy heiress, very nearly gone bankrupt and, according to one account, run away to Clifton in Bristol after being shot in the head by his wife's lover.

But on 18 February 1807 George was thinking how lucky he was. That day a gale had blown up and, as barrack-master at Great Yarmouth, he had come down to the shore with many others to watch the inevitable. As the storm raged, ships out to sea foundered whilst those in the supposed safety of the shallows broke their lines or fouled their anchors and were driven towards the beach. One particular horror that day was the fate of the gun-brig *Snipe*. She had become tangled in her anchorage forcing the captain to order her cables cut. Cut loose she drifted shorewards, her crew, some Napoleonic prisoners, and many women and children, still aboard. Near the pier, just 60 yards from where Manby stood, she then foundered in the huge waves and those on the beach heard the screams of the dying as they were thrown in the water. Sixty-seven people were lost, just yards from safety. The following day another 147 bodies were recovered along the coast from other ships that had sunk in the same storm.

Such shipwrecks were then considered the dreadful but inevitable consequence of seafaring – but not by George Manby. As an eighteen-year-old boy he had, as a prank, fired a line from a mortar over Downham church and it occurred to him that were he to be able to fire a line to a sinking ship, that might provide an escape route for her otherwise stranded crew. Fortunately Manby had gained a good engineering background at the Royal Arsenal so knew what tools he needed, but the problem of firing a heavy line accurately out to sea proved a hard nut to crack. In early experiments the rope caught fire, then the huge acceleration of the shot from the mortar broke the link between it and the heavy rope. Chains split and ropes frayed until finally he hit upon using platted rawhide which had both the strength and the flexibility to hold fast.

Manby's first chance to use his rescue apparatus in earnest came almost exactly a year after the sinking of the *Snipe*. On 12 February 1808 the brig *Elizabeth* got into difficulties in a storm just 150 yards from the Great Yarmouth beach. The crew had lashed themselves to the rigging and all appeared lost. Then George Manby and his portable rescue apparatus appeared on the beach and a line was successfully fired to the brig. This was lashed onto the deck and a boat sent out along the line. All seven members of the crew were brought safely to land.

More rescues followed and the apparatus eventually came to the notice of the House of Commons where Manby was rewarded with £2,000, although only after a tussle with a Lieutenant Bell who claimed to have also invented the device (although his was fired from the ship – no easy task when it's sinking). As well as the money, Manby was rewarded with seeing the commissioning of the first organised coastal rescue system in the country – a series of 'mortar stations' using his apparatus. As such he has been claimed as the true inspiration for the RNLI.

Success led Manby to redouble his efforts rescuing those in peril on the sea. The problem with night rescues was seeing far enough in a storm to accurately fire a mortar at a ship, and so he invented the 'night-ball', a mortar-launched flare, the precursor of the modern 'starshell'. Not all his inventions were quite so well received however. Along the coast the shoremen had for centuries supplemented their livelihood with salvage from sinking or distressed vessels and the idea of no ships sinking frankly disturbed them. When Manby demonstrated an 'unsinkable' boat, which used floatation tanks to retain buoyancy, the crew sabotaged the demonstration by rocking it until it capsized, tipping them, and Manby (who couldn't swim), into the water.

Realising that discretion is the better part of valour, Manby moved his life-saving activities inland, developing an apparatus for saving those who had fallen through the ice before developing his final triumph, the *Extincteur*.

Having dealt with water and ice, the *Extincteur* was Manby's method for dealing with that other great life-taker – fire. His machine consisted of a pressurised three-gallon copper vessel filled with a solution of the fire-suppressant potassium carbonate which could be directed onto the fire through a nozzle. In short it was the world's earliest modern fire-extinguisher. He then backed up this life-saving work by becoming the first to call for the formation of a national fire brigade.

This greatest of the life-preservers was himself rather well preserved. He died at his home in Great Yarmouth in 1854 just ten days short of his 89th birthday. At the time, his legacy was considered to be his huge collection of Nelson memorabilia but in fact it was perhaps the lives of thousands saved from fire, ice and water by his ingenious mind.

Episode Thirty – Ouch!

What's the most important thing in an inventor's arsenal? A screwdriver? A soldering iron? A patent lawyer? No – it must surely be a sticking plaster. Everyone gets the odd cut and graze from time to time when fiddling around with things and the sticking plaster is what enables us to mop up and carry on after such minor disasters. It might seem like a humble thing but, of course, there's a great engineer behind it. When I say 'great engineer' you could also justifiably write that as 'bored executive and his catastrophically clumsy wife'. And some boys in woggles. All are true.

Earle Dickson had a head start in the bandaging game as he worked for Johnson and Johnson, the American medical company that had been an early pioneer of sterile surgical dressings, following Robert Wood Johnson's conversion to the idea after hearing a speech by Joseph Lister. Dickson was not an engineer however, nor was he working in product development; indeed in 1920 he was just a humble cotton buyer at the company's HQ in New Brunswick, New Jersey.

Earle lived nearby in Highland Park with his bride of three years, Josephine Frances Knight. It was a happy marriage, if a slightly bloody one. The problem was not anything sinister however, simply the fact that Josephine had come to married life without the sort of manual dexterity that makes tasks like cooking a happy experience. In fact Earle had noticed from early on that her hands were constantly covered in pieces of wadding and medical tape, covering up cuts and burns received in her daily battles with the dinner. Nor was dinner Josephine's only problem. Having been wounded the received practice was to wash it, cover it with a piece of gauze and cotton wadding (preferably purchased from Mssrs Johnson and Johnson) and then bind the dressing on with tape. This was not an easy procedure to do at home on your own, with one hand, whilst bleeding. Furthermore, the bulky dressings made subsequent culinary upsets more likely and in washing those further burns and cuts, the old dressing would get soggy and the tape would peel off.

So like any decent, concerned human, Earle began to experiment. What was needed was a way of easily dressing a wound using only one hand and in such as way that the dressing would remain in place and not precipitate further disasters. Having access to the raw materials at J&J he began to cut strips of medical tape and place down the middle a line of cotton gauze. To keep this clean and dust free until it was needed he then covered the tape with crinoline. Whenever Josephine needed a dressing, which was often, she could now simply cut a piece of tape, peel off the backing and apply it to the wound. Still not the easiest task with one hand, but a great improvement.

Clearly the improvement in Josephine's quality of life was more generally noticeable too as Earle's friends at the office suggested he take his brilliant idea to his boss James Wood Johnson. Johnson liked the idea of a dressing you could apply yourself and agreed to put it into production. It would be nice to say that sales immediately took off, but sadly they didn't. In the first place the 'Band-Aid' plasters, as they were called, had to be made by hand, a slow and rather expensive job. Secondly they came in a roll three inches wide and eighteen inches long which proved a little cumbersome. Thirdly, no one knew they existed.

The answer to the first two problems came in 1924 when the company introduced machinery to make the plasters and decided to reduce the dimensions to something which made them more appropriate for small scrapes rather than major amputations.

The third problem proved more tricky. Earle's target audience had been his wife and whilst undoubtedly anyone working in a kitchen would welcome the odd plaster, in truth most people were not as clumsy as Josephine. What was needed was an audience who regularly fell over, cut themselves, scraped their knees and generally returned home of an evening looking like they'd been dragged through a hedge backwards. In other words, small boys. Distributing plasters to the small boys of America would be no mean feat however. Fortunately there was an organisation ready to take on just such a job. In a stroke of marketing genius, Johnson and Johnson sent free packs of sticking plasters to the organisers of Boy Scout troops. Soon, boys across America were traipsing home from their meetings, neatly bandaged up with Band-Aids and able to tell their parents of both their daring exploits and the new cutting-edge treatment for their cuts. This did the trick. By the time Earle Dickson died, in 1961, having risen to the heights of Vice-President of the company, it was making $30 million a year from his invention.

Episode Thirty-One – When in Doubt ...

Genetic engineering gets a bit of a bad press with talk of scientists 'playing God' and 'Frankenstein foods' and so forth. But in truth this has always been an area of engineering where the precautionary principle that, in the absence of scientific consensus, the burden of proof lies in showing that an intended action is not harmful, has held sway. This is largely thanks to the selfless behaviour of one man.

In early 1971 Paul Berg (1926–) and his team at Stanford University in California were working on the very edges of biological science. Following the pioneering work of Crick and Watson identifying the DNA helix, they were now looking at ways of splicing together sections of DNA from different organisms and then using a virus to inject the resulting genome into a living cell – a technique known today as Recombinant DNA (rDNA). This held out the prospect of identifying the effects of individual genes and perhaps later taking particular beneficial genes from one creature and inserting them into another. It was heady stuff – a chance to begin taking apart the instruction set of life and reorganising it.

During the 1960s a lot of work had been done using the bacterium every handwash manufacturer loves to hate, E. coli, and the virus 'Lambda' which infects it. What had not been attempted was expanding this research into studying mammalian cells and the viruses that might be used to insert new genetic material into them. By the early 1970s this was the area that was starting to interest Paul Berg. He wondered whether it might be possible to use the Simian virus 40 (SV40) to carry novel DNA sequences into mammalian cells.

The problem with SV40 was that it was tiny – only 5,000 base pairs long, encoding just five genes, so he realised he would have to tinker with it a bit to give it the ability to pick up other genes and carry them into a living cell. To achieve this he decided to try to splice together SV40 DNA with a DNA fragment that could replicate independently of a cell's genome (a plasmid), constructed from his old friend the lambda virus and three E. coli genes. This recombinant DNA would then be inserted into living E. coli cells.

It was this exciting development that Berg's research assistant Janet Metz found herself explaining to other pioneering genetic engineers at the Cold Spring Harbour Laboratory Tumor Virus Workshop, Long Island, in 1971. But the reaction to this brilliant work was perhaps not what she had expected. Robert Pollack, who worked at Cold Spring Harbour as a microbiologist, made a startling observation. The agent they were using, SV40, was a tumour-causing virus, which they had now given the ability to splice itself into the genome of E. coli. E. coli was, in turn, a very common bacterium resident in the intestines of just about every human alive. In an urgent call to Berg, Pollack asked 'What if the E. coli in your lab escaped into the environment and into people? ... We're in a pre-Hiroshima situation. It would be a real disaster if one of the agents now being handled in research should in fact be a real human cancer agent.' Could they have created a Frankenstein bacterium – a common human parasite now with the deadly ability to spread a cancer plague?

You'll be delighted to know that, in fact they hadn't, but the deadly possibilities of recombinant DNA had suddenly become clear. So what was to be done? Berg was initially concerned that Pollack was overreacting, but decided to canvass more opinion. Many argued that the work was so vital that risks were worthwhile, others that it was a matter for ethicists not scientists. One group, led by Nobel Laureate James Watson, claimed that the technique posed no particular threat and that 'overregulating' research would be more dangerous.

It was a delicate situation. The first steps of Berg's work had been completed – the rDNA had been painstakingly created. All that remained was to insert it into a living cell. But what would result? A miracle or a monster or just something interesting in between? With the certainty that executing the last step in the experiment would be Nobel-worthy stuff, many would have carried on regardless, leaving the theoretical problems for others, but uniquely Berg chose not to. On 26 July 1974 he published an open letter in *Science* calling for a voluntary moratorium on some areas of rDNA research (including his own) until the risks were better understood.

This was followed up the next year at the Conference on Biohazards in Biological Research (now known as the Asilomar I Conference), held at the Asilomar Conference Centre, Pacific Grove, California. Here genetic engineers openly discussed the possible outcomes of their research, bringing the subject to public and government attention and leading to the laying down of strict guidelines on the creation and use of rDNA. In the meantime another team had successfully spliced a gene from a toad into an E. coli bacterium, effectively stealing Berg's thunder.

But Berg's insistence on being an ethical engineer did bear fruit, ushering in a new era of openness in science. Far from restricting research, it also brought the ideas and terminology of genetic engineering into the public domain where they belong. It has not made life easy for genetic engineers but it has been the right decision. Paul Berg was finally awarded his Nobel Prize in 1980.

Episode Thirty-Two – Engineering Extinction

We're very good at making things go extinct – dodos being perhaps our most famous victim – although contrary to popular belief we didn't eat them to oblivion; in fact they tasted horrible. But this loss was, at least in our modern minds, an 'accident' and we didn't really mean it to happen. But what about actually deliberately engineering an extinction?

Alarming as this might sound we have made one concerted effort to eliminate an entire species and the story of how that went is perhaps a cautionary tale for our time. But why would we want to make a species extinct? Because it's an even bigger killer than us. Since its emergence in Africa some 10,000 years ago smallpox has killed more people than every war in history put together. The smallpox virus – Variola – only lives in humans, spread by airborne droplets or concealed on the bodies and clothes of victims, particularly in the thousands of burning, sore scabs that erupt on the skin. Within days of the scabs appearing there is profuse internal bleeding, vomiting of blood, and the sloughing of large pieces of dead skin. Those that survive never catch the disease again, but then roughly half of those who catch it don't survive. Within a fortnight they're dead.

Humans are, of course, resourceful creatures, and it wasn't long before we realised that it would be useful to have a killer like this on our side. So engineering Variola as a bio-weapon began surprisingly early, in the twelfth century, when the Mongols catapulted infected corpses into besieged cities. In the New World the disease, spread on infected blankets, was more feared than the Conquistadors – a fact used by the invaders to their advantage. The problem was, no one knew exactly how smallpox was spread, how it killed and how to prevent it, making it a rather unreliable ally.

It was the British physician Edward Jenner who made the breakthrough when he noticed in 1790 that milkmaids infected with the non-lethal but related cowpox developed a complete immunity to smallpox and this led to his discovery of vaccination – the word coming from the Latin for cow, 'vacca'. Even before that the Chinese had for some time been using quills to blow dried smallpox scabs up the nose in an attempt to inoculate their citizens, with some, if not complete, success. But Jenner's vaccination system was perhaps the first time in history that science had made a major impact on one of humankind's greatest adversaries. Sadly his idea was slow to catch on. Despite the willingness of many nobles, including Catherine the Great of Russia, to be vaccinated, his cure could not be produced in sufficiently large quantities to inoculate the entire population of Britain, let alone the world. So the steady, deadly spread of smallpox continued, through the nineteenth century and into the twentieth when it was still killing two million people worldwide a year.

What should have finally put an end to its reign of terror came in 1966, not in the form of a sudden scientific advance, but simply a resolution. That year the World Health Organisation decided to eradicate smallpox from the face of the earth. There followed a massive detective investigation in which teams of scientists scoured the planet for outbreaks of the disease, inoculated millions of people who were in danger of infection and finally brought the virus to its knees. The last known case of smallpox in the wild was located in Somalia on 26 October 1977. The victim survived.

But this was not quite the end of the road for smallpox. Although eradicated in the wild, the virus did still exist, frozen, in a few laboratories. In 1978 these numbered four, when Variola made its final bid for freedom. Janet Parker was a medical photographer working on the floor above a smallpox laboratory in Birmingham in the West Midlands when she contracted the disease. How the virus escaped has still never been explained but in the ensuing panic over 300 people who might have had contact with her were inoculated. Only one, her mother, contracted the disease and she survived. Janet was not so lucky. She remains the last known victim of smallpox. Several days after her death the director of the laboratory responsible committed suicide.

This tragedy finally brought smallpox to the condemned cell. Remaining stocks of the virus were moved to two facilities in the USA and Russia. From here the WHO decided to take the last step and destroy the disease entirely. It was a momentous decision as it would be the first time in history that humankind had deliberately made a species extinct. By a unanimous vote the committee decided that on 30 June 1993 the remaining vials of smallpox would be placed simultaneously in autoclaves in the USA and Russia and heated to 248 degrees Fahrenheit for 45 minutes. The dead and denatured viruses would then be incinerated.

But at the last minute we baulked. What might we lose if we destroyed a natural organism entirely? Might it be useful to the military? Or bio-engineers? Or medical researchers? Execution was postponed and the only engineered extinction put off, indefinitely. So smallpox is still with us today.

VII

In which we make a spot of dinner.

Episode Thirty-Three – For Your Convenience

Today eating at home is only marginally more effort than going out. Meals can be pre-prepared, sealed, preserved, packaged and delivered. All we need to do is take it home and put it in the ultimate modern cooker, the microwave, and within a couple of minutes a dinner, or something approximating to it, is ready.

Nor are we limited in which types of convenience meal we eat. Two hundred years ago what you ate, assuming you could afford to eat, was dictated by where you lived and what time of year it was. As the poet John Gay put it in 1716 as he walked through the food markets of his day:

Successive Crys the Seasons Change declare,
And mark the Monthly Progress of the Year.

A modern supermarket serves all foods at just about all times of year – for your convenience. It is as though the seasons that John Gay was so fond of have been made to stand still. Which strangely was the slogan of the product that started all this off.

It was in February 1809 that the *Courier de l'Europe* described Nicolas Appert's invention as 'making the seasons stand still'. And it was right, thanks to a short but forward-thinking megalomaniac called Napoleon.

At the turn of the nineteenth century Europe was in turmoil largely because of one man. Napoleon was a brilliant general and an astute politician but his overwhelming desire to conquer the world had led him into a spot of bother. He needed to give his army an 'edge' and his thoughts turned to food. Fighting at that time was a strange affair as the availability of food for men and horses meant there was an 'off season' in winter when most armies packed up and went home. What Napoleon needed was a way of preserving food grown in the summer, so his troops could eat it, and keep fighting, in the winter. No one would expect that. As he so famously put it – 'an army marches on its stomach'.

So he set up a competition with a magnificent 12,000-franc prize which was won, in 1809, by our friend Nicolas Appert. His method was simple but effective. By boiling the food first and then sealing it in glass jars he sterilised and protected it. It would then stay fresh until the jar was open. Napoleon was very happy. So was Appert who proudly called his method of preserving in glass jars 'canning'. Which was odd.

A year later, perhaps inspired by the name, the American Peter Durand patented a similar technique but using tin cans instead of glass. He called this process 'canning', which made a lot more sense. Nineteen years later the first tin canning factories opened in the USA and the boom in preserved foods began.

Now, Durand had been very clever to invent the tin can but he had forgotten something and it's thanks to another American that it was solved. Durand forgot to mention how you open these cans. The standard method was using a hammer and chisel, a clumsy technique that made your food taste of tools and your tools smell of food. But this clearly didn't bother too many people as the solution, the can opener, was only invented a staggering 48 years later by Ezra Warner of Waterbury, Connecticut.

Of course there's more than one way to skin a cat, not that that is a very appropriate idiom for an article on food, but with canning in the bag what today we might call the 'food technologists' started looking around for other ways to halt the seasons. And so we must allow a century to pass and follow another American pioneer north to Labrador. Clarence Birdseye was one of the most prolific inventors of his day, clocking up over 300 patents in his lifetime. The early career of this hero of food preservation was not exactly savoury however, as he paid his way through college by trapping rats to sell to researchers at the University of Columbia. From 1912 to 1915 he expanded his operations from vermin procurement to become a fully-fledged fur trader in Labrador, Canada, finding along the way a wife, Eleanor, and learning a great deal about his icy environment. In particular he had noted how animal carcasses frozen in the depth of winter survived much better than those frozen in spring or autumn and this set him thinking. After rather annoying Eleanor by repeatedly filling all her washbasins with cabbage leaves and brine and then leaving them to freeze solid outdoors, he finally worked out what was happening.

What Clarence had discovered was fast freezing. When foods were slowly frozen, large ice crystals formed which damaged the cell structure making the defrosted cabbage leaves less appetising (if defrosted cabbage leaves can ever be considered to be anything else). Fast freezing however led to the creation of tiny ice crystals which preserved the structure of the food and the resulting defrosted cabbage leaves were every bit as delicious as the moment they were frozen. Of course the local Inuit peoples had known all about fast freezing for centuries, as, to be fair, Clarence pointed out, but Clarence was the first to work out why it worked and then apply it to the mass production of frozen foods. The rest is convenience food history and Clarence went on to make his name and fortune freezing everything from peas to cod (but not rats thankfully). And so with canning and freezing the TV dinner went from being a food technologist's dream to a processed reality.

But there's one engineering marvel still missing – how to heat up this feast fit for a king? Which brings us onto the subject of chocolate. Put chocolate in a warm place and before long it turns from solid into liquid. This fact is every mother's distress and every dry cleaner's delight.

It was also known to Dr Percy Spencer, a scientist working at the Raytheon Corporation laboratory in the USA in 1946. He was tinkering with a new device called a magnetron when he noticed the chocolate bar in his pocket had melted and ruined his shirt. Now this was unusual – very unusual. There was nothing hot around.

Spencer eyed the magnetron with suspicion. He tried putting popcorn kernels near it – and sure enough they popped. The next day he put a fresh egg by it and was surprised to see the egg begin to vibrate and rattle before exploding and showering him with omelette. By this point he was getting through a lot of washing, but he was also getting the point. Microwave radiation from the magnetron was exciting the water molecules in the food and cooking things – incredibly quickly. Spencer decided to put a magnetron in a metal box to reflect and concentrate the waves and the microwave oven, and hence the true convenience meal, was born.

Ketchup anyone?

Episode Thirty-Four – Stephenson's Rocket

'That's not very eccentric,' I hear you cry, and indeed were I to write about Stephenson's famous locomotive to an audience of engineers such as yourselves, for whom such things are bread and butter, you'd be quite right. But I'm not. Instead I'm talking vegetables. Not the leafy green 'rocket' per se (that was just to get your attention) but a nevertheless extraordinary piece of vegetable engineering for which one of our greatest nineteenth-century engineers can claim the credit.

When George Stephenson retired from the railway business to his Tapton House estate in Chesterfield he was determined to set his mind to new problems. The first was ensuring that no one would forget his contribution to the industrial age, which he achieved by regularly reminding everyone and anyone in earshot of his humble background, lack of formal education and years of struggle promoting railways. The second problem was another area of the new Victorian world that clearly needed 'improving' – horticulture.

Stephenson brought an engineer's mind to the task of bettering what was, in these pre-Darwinian days, generally considered to be God's creation. His interest in growing exceptional vegetables had probably been kindled during his years at Killingworth where he had been appointed as brakesman to the West Moor colliery engine in 1804. Here he had delighted in growing enormous cauliflowers and cabbages to win local horticultural competitions.

Now rich and retired he looked to tame more exotic fruits, in particular melons and pineapples. The problem of growing large and perfectly formed melons was solved by his invention of a form of wire basket to hold the fruit suspended with the stalk not under tension, ensuring the melon grew round and clean and was not choked of nourishment from the strain it put on its own stalk. Pineapples also succumbed to his relentless engineering genius thanks to the construction of ten hot-water heated forcing houses and a range of pineries, one 140 foot in length. If his pineapples did not ever quite live up to his boast that he would grow them 'as large as pumpkins' he did at least eventually beat his arch rival, the Duke of Devonshire, in competition, although the winning 'Queen pines' were only produced a year after his death.

However, his most vexatious vegetable was an altogether humbler thing – the cucumber. Not that Stephenson had any trouble growing them to a magnificent size in his greenhouses, aided by his careful selection of manure and the lavish attentions of his legion of gardeners. What bothered Stephenson was their shape. Cucumbers, he discovered, would not grow straight and, being an engineer who was fond of straight lines, this bothered him.

Fixing the problem was no simple matter however and even he was at first bemused. Initially he tried his melon trick, suspending the cucumbers on wires in the hope that gravity could play a steadying hand in their development. When this failed he tried growing them on trays using various mechanical props to counter any emerging curve before it became too severe. This also failed.

Never a man to give up, as he would have told you by now had you been in audible range, he next turned to the application of heat and light in the hope that the cucumbers were bending towards or away from the source of one or the other and so by applying the same to the other side, the curve could be corrected. Had this been successful there might today be a whole career in manoeuvring paraffin lamps and torches around individual cucumbers. Fortunately it did not.

In the end it was not science that would tame Stephenson's cucumbers, but the application of brute force. If they would not grow straight in any natural environment, he reasoned, then perhaps he could grow them in some kind of mechanical mould that would prevent them growing any other way. Having dashed off a quick design he sent his idea to a glassworks in Newcastle where it produced for him a series of thick-walled glass tubes open at both ends and tapered towards one. When the young cucumber was placed inside one of these straight tubes it could do nothing but grow straight, or so Stephenson hoped.

And the application of this exceptional engineering mind to this most peculiar of problems did indeed work. That other great improver Samuel Smiles notes in his *Life of George Stephenson* that one afternoon the great man finally returned triumphant from the garden wielding one of his dead-straight cucumbers and announced to his gathered house guests 'I think I have bothered them now!'

Episode Thirty-Five – Hard Tack

It's not always been easy being in the Royal Navy. In 1797 naval ratings got their first pay rise in 139 years and even then the money was, frankly, disappointing. Then there was the savage corporal (and sometimes capital) punishment dolled out by the officers. Then, assuming you shrugged off the penury and beatings, there was the food.

Sailors in the nineteenth century didn't even get their 'daily bread'. Bread was simple enough to prepare but it did require a slightly unusual ingredient – live yeast. Unlike Her Majesty's jolly tars, yeast wasn't fond of long sea voyages and, when press-ganged, had a habit of dying, just to annoy the baker. So the answer for most on board was the truly gruelling hard tack biscuit.

Hard tack was an even simpler food, made from flour and water and, if you were lucky, a bit of salt. This glutinous paste was shaped into rounds or squares and baked hard. Then it was baked hard again. This was how poorer landlubbers came across it, but for sailors it was considerably worse. Ships are damp places and damp food tends to go off so it was essential that the hard tack was absolutely bone dry when it went into the hold. To ensure this the navy baked its tack not twice, not three times but four times, creating in the process a snack with many of the physical properties of cement. It also liked to bake early, often preparing the biscuits a full six months before a ship set sail. In this time the only creature more tenacious than the British sailor made its own heroic inroads into the stock – the biscuit weevil.

Clearly a change was needed and fortunately, in 1845, Henry James Jones came on the scene. Jones, a Bristol baker, had been puzzling over the problem of how to make bread without yeast. It was the need to keep yeast alive that prevented everyone from housewives to sailors from making their own bread and, after much experimentation, he had come up with a method of combining ordinary flour with tartaric acid, bicarbonate of soda, a dash of sugar and a pinch of salt. This seemed to do the trick. Here was a flour that could be mixed with just water to make a bread dough that would rise like a yeast dough. And so the rest should have been history.

In fact the new 'self-raising flour' was a hit at home and Jones was a great promoter of his flour which he sold in 'signed' bright yellow bags with blue lettering. Patents were duly granted and the judicious mailing of samples of the new flour to the great and good soon brought a Royal Warrant along with hearty recommendations from *The Lancet* and even Florence Nightingale. Naval men also sent their compliments, including the captain of the SS *Great Britain*, and the chef aboard the Royal Yacht.

But one door remained firmly closed – the Admiralty's. Their Lordships were not known for taking hasty decisions – remember the 139 years without a pay rise – and the chance to put fresh bread on the mess table was not going to make them move any faster. Ten years passed without an Admiralty decision on the patent flour in which time Jones tried everything to persuade them. In 1846 he invented an on-board bread machine designed to bake his self-raising bread which he demonstrated to the Admirals. They dithered for a year before demanding the machine be sent to Woolwich for a full trial. When no word came back Jones wrote again receiving a curt reply telling him his machine had been broken up for scrap. A legal case followed before Jones recouped any of his costs.

The following year the report into the testing of the machine finally emerged and was wholly positive. This spurred the Admiralty into action and six days later it sent Jones a letter saying it was rejecting both his flour and his machine. Six more years passed before Jones tried again. In one last bid he gathered together all his correspondence with the Admiralty along with all the many recommendations he had received, including one from the Director-General of Naval Hospitals. These he published in a pamphlet which he sent to every single member of parliament, with a note saying:

that a grave responsibility would rest upon himself if he did not make this attempt.

Their Lordships, suddenly finding themselves in the uncomfortable glare of adverse publicity, saw the light. Within a month Jones' self-raising flour was on the provisioning lists of every ship in the Royal Navy. However, just in case the sailors thought all their birthdays had come at once, its use was restricted to Sundays. For the other six days it remained hard tack and weevils as usual.

Episode Thirty-Six – A Slippery Customer

Lots of the best ideas in engineering come about through luck. That's not to say that engineers simply wander around bumping into things hoping that something interesting will happen, but it is true that unintended consequences, as often as intended ones, can lead to a breakthrough. It is the job of the discoverer or inventor to simply be open to the possibility and on the lookout for the unusual. As Louis Pasteur put it, 'chance only favours the mind which is prepared'. Just over a century ago a man with just such a mind was born in the town of New Carlisle, Ohio, and his story shows just how far an unintended consequence can go.

Roy Plunkett's contribution to the world could, were it not for one happy accident, have seemed to modern eyes rather environmentally unfriendly. In an era long before the dangers were fully understood he took a leading role in the introduction of tetraethyl lead to petrol as an anti-knocking agent and when he wasn't doing that, he was trying to create novel CFC refrigerants which we now know can leave an unfortunate hole in the ozone layer.

But it was thanks to CFCs that Plunkett had his great moment of discovery and one which has made the lives of many an engineer, heart patient and cook much the better for it. In 1938 Plunkett, a research chemist at the DuPont company, was investigating a toxic problem. Refrigerators at that date contained substances like ammonia which were all perfectly safe whilst enclosed in sealed systems but which had a nasty habit of leaking out and poisoning anyone nearby. As poisoning chefs is generally considered poor form he was charged with finding a safer alternative using chlorofluorocarbon compounds, work which would eventually lead to the synthesis of 'freon'. Had this been all then Plunkett's contribution might have been hailed by refrigerator users but decried by environmentalists, but it was whilst working with CFCs that he made another, more unusual discovery.

Plunkett had produced a quantity of the gas tetraflouroethylene (TFE), a simple fluorocarbon used in preparing polymers, and, this being a shade volatile, he was sensibly keeping it in small steel cylinders cooled with solid carbon dioxide. The odd thing was that when he went to tap the TFE in some of the cylinders a few days later, nothing came out. Initially he assumed that the gas might have escaped, or perhaps the pressure valves on the cylinders had stuck shut. There was only one way to find out. Carefully taking one of the cylinders outside he placed it behind a specially built shield and cut it open. Inside there appeared to be nothing, which was odd, as the cylinder weighed just as much as when it had first been filled.

Then he noticed a white powder lining the inside of the container. The TFE had accidentally polymerised to form polytetrafluoroethylene (PTFE), catalysed by the iron in the cylinder walls. This was worth a bit of further investigation even if it wasn't exactly what DuPont had asked him to do so he took away the powder for analysis. In his notebook he recorded,

it is thermoplastic, melts at a temperature approaching red heat, boils away without residue, is insoluble in hot and cold water, insoluble in all petroleum solvents, alcohols, ethers, all acids and alkalis ...

To put it another way it was almost completely inert. Nor did this powder's peculiarities end there. It also had an extremely high electrical resistance but more interestingly than that, it had an incredibly low coefficient of friction, in fact the second lowest of any known solid material, the lowest being 'diamond-like carbon'. This meant that nothing stuck to it.

So what use is a substance that nothing reacts with and nothing sticks to? DuPont, who christened the substance 'Teflon', could think of thousands. Thanks to its inertness it is today used to line valves and pipes carrying highly volatile substances, its dielectric properties make it an excellent insulator in cables and circuit boards, whilst in powdered form it can be used as a pyrotechnic oxidiser in decoy flares used to confuse heat-seeking missiles.

But it is its slipperiness which has really made its name. Sprayed on furnishing fabrics and carpets it makes them highly resistant to dirt whilst on a larger scale, sprayed on the whole of the Millennium Dome, it helps to keep the building a sparkling white(ish). The whole of the Sydney Opera House sits on Teflon pads which allow it to respond to thermal expansion and even Arctic expeditions benefit from its use, on sledge runners. As a slippery coating it eases the path of deadly armour-piercing bullets but can also form vascular grafts which prevent clots forming and hence save lives. Even pet shops have Teflon to thank as it makes an ideal coating for insect cages, as even ants can't climb up it. It is also the only solid substance that can defeat a gecko.

But the most famous application of this unexpected result came about in 1954 when French engineer Marc Grégoire was thinking about how Teflon might stop his fishing tackle getting tangled. His wife suggested that a more useful problem to tackle might be stopping food sticking to her cooking pans. So Grégoire created the first non-stick resin for cookery and from there the slippery stuff found its way into every home.

VIII

I ORTALITY

In which we build things, move things, hide things and watch them fall down.

Episode Thirty-Seven – The Hypogeum

There's always a danger in engineering that it is seen as something purely modern or at least recent. Ask people what image swims into view when you whisper 'engineering' in their ear and they'll probably say something about Brunel, bridges or jet engines. Or they might simply ask you to stop behaving in such a peculiar way.

But engineering is, of course, part of the bedrock of civilisation and has hence been with us as long as civilisation itself. And so it's back to those beginnings and into that bedrock that I thought I'd venture.

Everybody knows that the first engineers were Egyptian – or Sumerians, or Chinese. Or were they perhaps Maltese? It was in 1902 that a group of builders digging a well for a new housing development at Paola in the Hal Saflieni region of Malta fell through into what they assumed was one of the natural caves that riddle the limestone foundations of the island. On closer inspection the cave clearly showed signs of human use which was a shade inconvenient so the builders decided to keep quiet and just use the hole as a useful dump for building rubble.

For two years the builders valiantly kept their naughty secret but eventually news leaked out to Father Manuel Magri, a local priest with an interest in Maltese archaeology who was asked to investigate on behalf of the

Museums Committee. What he uncovered, having removed the building rubble, was one of the most remarkable prehistoric sites in the world – the Hypogeum – consisting of a complex of twenty rock-cut chambers, each connected to the next by sinuous passageways, descending in three levels and filled with human remains.

Malta and its neighbouring island of Gozo were already well known for their above-ground prehistoric temples, but here was the process in reverse – a temple created by removing material, being cut into the soft limestone bedrock, initially modifying the shape of natural caves and then cutting whole new rooms from the living rock itself.

The site covers over 500 square metres and descends into the ground in three distinct levels. At the highest level the remains of over 7,000 people were found, suggesting this was originally a communal burial place. The level below is more complex and seems to have been constructed from scratch. The main chamber here is almost circular in shape and was originally painted red. Its entrance ways, some of which are real and some false, were carved to mimic the trilithons (two upright stones with a third across the top) which form the doorways of the surface temples on the island. In this room lay a remarkable figure, known today as the Sleeping Lady – a small statuette of a woman with a tiny head and very large body, reclining on a couch. What she represents however is a mystery.

From this main chamber other rooms lead off, whose names are, of course, not original but were given to them by the archaeologists who excavated them. The Oracle Room with its strange acoustic, the Decorated Room with its carved stone 'handprint', the two-metre hole known as the Snake Pit and the 'Holy of Holies', a small room with a coffered ceiling containing a window or 'porthole' in a trilithon framed by two larger trilithons. A third level exists below this but no human remains have ever been found there. Archaeologists have suggested that it may have provided a storage area for food but under a necropolis is not where I'd choose to store my dinner.

But the greatest mystery about the Hypogeum was its date. It was carbon dating, invented nearly 50 years after the discovery of the site, that finally answered the question and in the process changed both the history of Malta and the world beyond. Dates taken from organic remains at the Hypogeum (in the form of bones) and some ancient wood samples from the above-ground temples of the same style as the Hypogeum, revealed that these structures were not Bronze or Iron Age, as might be expected if the technology to make them had diffused there from the Near East, but Neolithic, the earliest dates coming from a thousand years before the Egyptian Great Pyramid was even begun. In fact one of the temple sites, at Ġgantija, proved to be the oldest free-standing stone building known anywhere on earth.

The discovery of the enormous antiquity of the Hypogeum, which was begun between 3600 BC and 3300 BC (for the upper levels), finally being finished around 2500 BC for the lowest level, also had startling implications as to the sophistication of those early engineers who made it. Coming from the Stone Age, these beautifully carved caverns must have been cut without the help of metals using just stone, bone and antler tools. Furthermore, as the flint and greenstone used in these tools cannot be found on the island, an extensive foreign trading system must already have existed to provide them. Before the Egyptian Old Kingdom had even begun, here was a culture with a sophisticated architectural knowledge whose influence spread beyond its own shores. So perhaps any engineer thinking of going on a pilgrimage to their roots should cancel that ticket to the Near East and head for Malta instead, where, five and a half millennia ago, it all began.

Episode Thirty-Eight – Cleopatra's Needle

Not far from the IET's London home in Savoy Place stands one of the city's most famous monuments – Cleopatra's Needle. An odd place, you might think, to find an ancient Egyptian obelisk, and the story of how it got there is no less unusual.

But to start at the beginning there are two things you should know about Cleopatra's Needle. Firstly it is only one of three such obelisks to bear that name (its pair being in Central Park NYC and another with the same moniker standing in the Place de la Concorde in Paris). Secondly none of them were built by Cleopatra. The London obelisk was commissioned by the Pharaoh Tuthmosis III some 1,400 years before Cleopatra was even born and was erected outside a temple in Heliopolis. A couple of hundred years later the Pharaoh Rameses II decided to have inscriptions carved on it, something he liked to do to any spare piece of stonework lying around in his kingdom which didn't already make mention of his great deeds.

It was only under Roman rule, in 12 BC, that the obelisks were taken to the city that had been until recently Cleopatra's capital, Alexandria. Here they were erected again outside the Sebasteum which Cleopatra had begun in honour of Mark Antony but which Augustus, after their deaths, finished in his own honour and renamed the Caesarium.

It was here that they stayed, although toppled by a series of earthquakes, until the nineteenth century. In 1801 the Ottoman viceroy of Egypt was looking around for something nice to give the British to thank them for Nelson's victory at the Battle of the Nile and Abercromby's at Alexandria, and he happened upon the fallen monument. It was an awkward gift however, weighing 187 tons and being, rather inconveniently, in Egypt, and the British government wouldn't pay to ship the huge monument to London. So it remained uncollected for 76 years.

Finally, in 1877, a wealthy English dermatologist, Sir William Wilson, agreed to stump up the eye-watering £10,000 needed to bring the prize to London. Electrical engineers amongst you might remember Wilson as the man who famously predicted in 1878 that 'When the Paris Exhibition closes, electric light will close with it and no more be heard of.' But let's forgive him that for now.

The question in 1877 was how to move this vast stone obelisk to London. Once the Needle was excavated from where it had fallen, engineer John Dixon was hired to construct a special barge for the purpose. Dixon came up with a cylindrical iron design 28 metres in length and nearly 5 metres wide. To stop the barge rolling over and over in high seas, Dixon fitted his design with two bilge keels, ballast consisting of railway track, and a small mast on which balancing sails could be hoisted. A rudder at the stern provided some limited manoeuvrability and a deckhouse was constructed from which the vessel's appointed captain, Henry Carter of P&O and his five crew, could control this unusual vessel. With the obelisk inside the barge, now christened *Cleopatra*, all that remained was to attach a line to Captain Booth's merchantman *Olga* which was due to leave for Newcastle-upon-Tyne with a cargo of grain and have her towed to England.

So far so good. Well not all that far really. On 14 October, the *Olga* and *Cleopatra* encountered gales in the Bay of Biscay. The cylindrical barge soon became unstable, rolling wildly in the building seas causing the ballast to shift. Carter and his men tried to move the ballast back but failed and when the balancing mast was lost he signalled for help. Captain Booth sent out a boat to rescue the crew but it was overwhelmed and all six drowned. Eventually he managed to get a line to the barge and took off her crew before cutting the cables and heading away, reporting in his log that the *Cleopatra* was now abandoned and sinking.

In fact, left to its own devices, Dixon's barge managed to ride out the storm rather well and four days later a Spanish fishing fleet was no doubt a shade surprised to find the huge iron cylinder bobbing quietly in the Bay. When they reported the find, the owners of the nearby steamer *Fitzmaurice*, out of Glasgow, seized their chance to make a few quid and towed the barge into Ferrol where they claimed a £5,000 salvage fee. After repairs and much wrangling, the fee was reduced to £2,000 and the paddle tug *Anglia* was dispatched to take the *Cleopatra* on the final leg of her journey.

She arrived on the Thames on 21 January 1878 and immediately parliamentarians began debating where the obelisk should be placed. You might think that they had already had 76 years to think about this but they remained unsure and the original plan to place it outside Parliament was eventually rejected in favour of a position on the Embankment. Cleopatra's Needle was finally unveiled there on 12 September 1878. Underneath the monument's pedestal was placed a 'time capsule' including a description of the eventful journey of the obelisk to London, no doubt with the fervent wishes of all engineers concerned that this particular granite monster would never have to be moved again.

Episode Thirty-Nine – Fame

Engineers don't always get the fame they deserve. Take poor old Maurice Koechlin. He was the real brains behind the armature that holds up the Statue of Liberty but the credit always goes to the sculptor Frédéric Bartholdi or perhaps the engineer he chose to help him, Gustave Eiffel. Eiffel of course would later repeat the trick with a certain tower that still bears his name despite the structural design being, once again, by the diminutive Koechlin.

So the question that's been bothering me is how can those who actually build great monuments get the recognition they deserve? How can they take their rightful place alongside the names of the sponsors and politicians whose names tend to smother such constructions? Isambard Kingdom Brunel was lucky enough to die famous enough for the words 'I. K. Brunel – Engineer, 1859' to be added in huge steel letters on each end of the Royal Albert Bridge over the Tamar. But only after his death. Most however become more rather than less obscure after shuffling off this mortal coil.

Except one. There is one great historical lesson for engineers everywhere on how to achieve your own slice of immortality, but you have to go back a long way to find it and it does involve a bit of cheating. Ptolemy II was an Egyptian pharaoh with a problem. His lovely, and relatively new, capital of Alexandria was on an almost totally flat coastline, making the harbour hard to locate from vessels out to sea. Not only that, those ships unwise enough to stray too close to the shore would find themselves in dangerous shoal waters, studded with reefs and sand banks where all but the most skilful might founder. Even with the harbour in sight, the most dangerous reef lay hidden across the entrance itself. Worse still, those unwise enough or simply unable to take the correct bearing might find themselves wrecked on the shores of the island called Pharos where the notorious inhabitants of the 'Port of Pirates' would easily pick them off.

What Ptolemy really needed then was a lighthouse. The man chosen for the job was Sostratus the Cnidian, whose previous work had included a 'hanging garden' in his home town of Cnidus in Caria and a clubhouse in Delphi where his fellow Cnidians could meet. For Alexandria he proposed something altogether grander. The great lighthouse was to be constructed in granite and limestone blocks faced with white marble. Its total height would be at least 120 metres – about the height of a modern 40-floor skyscraper. Erected on a solid base it would have three levels. The first level would be square and would house all the workers needed to fuel and operate the great light. The second floor was octagonal in shape, decorated with exquisite statuary and where the people of Alexandria might come to enjoy the breathtaking views. The third level was then circular and was crowned with an enormous reflector made of polished brass. Finally atop this stood a huge bronze statue of Poseidon, God of the Sea, leaning jauntily on his trident.

The light was designed to operate both day and night. During the day it was simple enough to just reflect the rays of the sun out to sea but at night something more was required. For this a circular shaft ran up the centre of the entire building, which enclosed the spiral staircase that gave access to the higher floors. Up the middle of this could be winched the piles of resinous acacia and tamarisk wood that provided the fuel for a great bonfire whose light was reflected far out to sea each evening by the great mirror. The location for the lighthouse was obvious, on the island whose name would come to stand for lighthouses everywhere – Pharos. Here it would be most conspicuous, not least to the wreckers and pirates who called the island home.

The building project took twelve years to complete and was said to have cost Ptolemy about 800 talents (a talent being 27 kg of gold). It was finally finished in the early years of the third century BC but this brought Sostratus his biggest problem. Obviously the lighthouse, a wonder of the world, would be dedicated to the Gods and to Ptolemy, the divine Pharaoh who had ordered its construction. But what about him?

According to Lucian of Samosata, Sostratus knew he couldn't compete with Ptolemy, but he also knew his lighthouse was built to last, indeed the Moorish traveller Yusuf Ibn al-Shaikh saw it still functioning in AD 1165, some 1,450 years later. So he carved his own inscription on the building:

Sostratus, son of Dexiphanes the Cnidian, dedicated this to the Saviour Gods, on behalf of all those who sail the seas.

But he then plastered over this and in the plaster wrote a great dedication to Ptolemy, and Ptolemy alone, making the Pharaoh jolly happy. Sostratus was thankfully a patient man as he would have to wait for his fame but he knew it would come. As the centuries passed, the plaster crumbled, consigning Ptolemy's name to the dust and leaving in its place the name of the monument's true builder, Sostratus the Cnidian – one engineer who was not going to be forgotten.

Episode Forty – Indiana Jones Eat Your Heart Out

In the West we like to celebrate our engineering wonders and have done so ever since Antipater of Sidon wrote down that rather famous list of his seven favourites. But not everyone likes to shout about their achievements and this has proved something of a double-edged sword for one of the greatest unsung builders in history, the Japanese emperor Nintoku.

Outside of Japan not a lot of people have heard of Nintoku's tomb which stands in the city of Sakai in the prefecture of Osaka and consists of a tree-covered keyhole-shaped earthen mound surrounded by three moats. According to the Nihon Shoki (the Chronicles of Japan which record the lives of the first thirteen emperors), Nintoku was the fourth son of the emperor Ojin, reigning from AD 313 to 399. This surprisingly long reign of some 86 years arouses suspicion, however, as does the chronicle's claim that he was 142 years old when he died. A Chinese chronicle, the Book of Song, which is probably a more reliable source for the period, records Nintoku sending diplomatic missions to China a century later in AD 421 and 425 but does as least suggest that Nintoku was a real emperor.

The Nihon Shoki also tells a story which is claimed by many Japanese scholars to mark a crucial turning point in Japanese history. It states that Nintoku looked out from his palace across the land and saw that there was no smoke rising from the houses as no one was cooking food due to extreme poverty. In response he abolished conscripted labour in Japan for three years to enable the people to concentrate on their own wellbeing during which time his clothes fell to rags and his palaces began to collapse. When the empress complained he showed her the smoke now rising from the homes of his people and told her that as his people were now cooking they were wealthy and if they were wealthy how could he be poor. This chimes rather neatly with the Confucian belief that rulers were placed on earth to guarantee, above all other things, the wellbeing of their people.

This legend may be all that remains of the memory of a reforming emperor, as beyond these references hardly anything survives to tell us of his rule. That is apart from this tomb. The mound is one of 92 large and small burial sites known as the Mozu Tumulus Cluster which lie within an area of about 16 square kilometres, the ten closest of which may hold the remains of Nintoku's royal entourage. These mounds are known as Kofun, and they give rise to the name of the whole proto-historic period of Japan's past, known as the 'Kofun Period', which lasted from *c.* AD 150 to AD 538.

To be fair this type of mound is not uncommon, indeed it is estimated that there are some 200,000 of them in various shapes – round, square or, as in the case of Nintoku's, keyhole-shaped. On average they measure up to a fairly impressive 50 metres in diameter and rise eight to ten metres high, but what is extraordinary about Nintoku's tomb is that it is much larger – vastly larger. Measuring 486 metres long and 25 metres high, it covers 32 hectares of ground and is thought to have taken 20 years to build with a workforce of around 800,000 men. Considering that the population of Japan in AD 400 is estimated to be only 1.5 million, this makes the tomb quite a sizeable national undertaking. So much for him abolishing conscription. In terms of its volume (above water level) it is the third largest mausoleum on earth after the tomb of the First Emperor of China near Xian (3,000,000 cubic metres) and the Great Pyramid of Khufu at Giza in Egypt (2,600,000 cubic metres).

Of the structure itself, what is currently known has been gleaned from visits in earlier centuries, photographic work and comparison with other non-royal tombs, as the Japanese Imperial Household does not allow visitors to even cross the outermost moat. Originally the entire key shape was clad in stone slabs, which must today lie beneath the trees which cover the site. It is estimated that around 26,000 tons of these would have been needed to cover the mound. Inside is probably a stone corridor leading to a burial chamber; indeed there are reports of an 'off the record' excavation taking place on the mound in 1872 following the accidental collapse of part of the structure, which confirms this. During that work it is claimed that a stone-lined chamber was exposed in the oblong end of the tomb which contained a stone sarcophagus. In this chamber was also found some iron armour and weapons, some gilt-bronze ornaments, a mirror and a sword, all of which are now preserved in the Boston Museum of Fine Arts in the USA.

What else remains inside is a mystery and the Imperial Household Agency has no intention of allowing anyone to find out any time soon. This undoubtedly fulfils the Agency's requirement to maintain the 'quiet and dignity' of the tomb but has robbed Nintoku of his rightful place amongst the great builders of antiquity. It has however left him in perhaps the most tantalising unexcavated site on earth.

Episode Forty-One – Hidden Treasures

No piece of engineering appeals more to the schoolchild in us than the secret tomb. If you haven't ever imagined yourself as a latter-day Indiana Jones, dashing down concealed corridors and dodging ancient booby-traps to discover a priceless treasure, then you really should, not least because there is actually a grain of truth to the movie legend.

By AD 900 the city of Palenque in what is today southern Mexico was a ruin, covered in forest and slipping from memory. It would be over 650 years before any outsider would see the remains and another 250 years of occasional reports before the site came to the attention of archaeologists. Following the publication of *Descriptions of the Ruins of an Ancient City discovered near Palenque*, in 1822 interest in these mysterious Central American lost cities began to grow in the West and in 1839 Frederick Catherwood, an architect and draughtsman, and John Lloyd Stephens, a diplomat and writer, finally travelled to the region where they glimpsed a pyramid poking through the thick undergrowth of the Chiapas foothills. Setting up camp in the ruins, Stephens and Catherwood went about recording what at first appeared to be an isolated temple but which they soon realised was an entire city choked by almost impenetrable rainforest.

The temple they were camping in is known now as the Temple of the Inscriptions and was begun in around AD 675 when Palenque was the capital of possibly the largest and most important kingdom in the Mayan world. A 23-metre-high stepped pyramid with a temple structure on the top, the building was decorated with the second longest inscription known from the Mayan world. Not that Stephens or Catherwood could read the text. Instead they simply drew what they saw – a beautiful, empty temple in a long-abandoned city with no sign or clue to where the people had gone, what fate had befallen them or where they were buried.

Had they been able to read it they would have learned that the inscription records 180 years of Mayan history at the city, and in particular details the life and achievements of the man who commissioned the temple, K'inich Janaab' Pakal, who is known today as Pacal the Great – the man who had revived the city's fortunes. What they never guessed was that he was still there. Nor had they any reason to. Since the rediscovery of Mayan civilisation it had been widely assumed that the stepped pyramids found in their cities were simply elaborate supports for the temples on their summits. There was no evidence for tombs, no evidence really for any of the people who must have once lived in these places. Just empty, ruined stone buildings.

And this was still the received wisdom when Mexican archaeologist Alberto Ruz Lhuillier visited the site in 1949. However, standing where

Stephens and Catherwood had stood 110 years earlier he noticed a double row of stone plugs set into one of the slabs on the sanctuary floor which he guessed marked the lifting points where the stone had been lowered into position, implying there might be something beneath it.

Removing the stone he found he was right. Beneath lay a vaulted passage so full of rubble that it took four digging seasons to clear and it was 1952 before his team arrived at the bottom. Here they found their way blocked by a wall next to which stood a stone box that contained pottery jars, shells filled with the red pigment cinnabar (mercury sulphide), beads, jade earplugs and a solitary pearl. Removing the wall blocking their path they next came to a more grisly discovery: a chamber containing the skeletons of six human sacrifices, beyond which lay another large block of stone.

On the other side lay something unheard of in Mayan archaeology – a royal tomb in the heart of a pyramid. Inside, a large slab covered an elaborately decorated sarcophagus, standing on six short piers, beneath which lay pottery food dishes and two life-size stucco human heads. The lid of the sarcophagus was decorated with one of the most important scenes on any Mayan monument, a depiction of its owner, Pacal the Great himself, who is shown falling into the netherworld whose jaws gape open to receive him.

Beneath the lid lay the body of Pacal, sprinkled with mercury sulphide and surrounded by over 700 jade items. He wore a jade diadem, a net skirt made up of jade pieces held together with gold wire, necklaces, pectoral decorations, rings, bracelets and ear-flares. Over his face lay a mosaic jade mask and at his feet rested two jade statuettes. In his hands he held a jade cube and sphere whose significance is still a mystery.

It was the find of the decade and one of the greatest ever Mayan discoveries. Here at last was one of the inhabitants of these lost cities, and one of their greatest – hidden for nearly 1,300 years, surrounded by his treasures. It was the sort of discovery you only find in movies and even today, with perhaps only five per cent of this 65-square kilometre city excavated, Pacal's kingdom holds on to many more secrets.

Episode Forty-Two – Coade Stone

London is full of lions, a fact that is not as alarming as it might first appear, as most of these are made of stone. The most famous, of course, lounge idly in Trafalgar Square but there are two, one by Westminster Bridge and one at Twickenham, often now overlooked, that tell the story of a remarkable woman and her even more remarkable material.

Eleanor Coade was all her adult life known as 'Mrs Coade' despite the fact that she never married and that in itself is a clue to her unusual position in society. The term 'Mrs' was applied as a matter of courtesy because Eleanor Coade was a businesswoman. After her father had died bankrupt she had joined Daniel Pincot in a business at Narrow Wall, Lambeth, making an exciting new material – artificial stone. It was a great time to be in the building trade as the finest Georgian architects were then at work, including Robert Adam and John Nash. The demand for their work was huge and the demands this placed on stonemasons were even larger. The great aristocratic families wanted new Neoclassical houses with Neoclassical ornaments but making these by hand was a slow and expensive job. Fortunately Eleanor had the answer.

Coade stone, which she called 'Lythodipyra' from the Greek 'twice-fired stone' had the appearance of the greyish-yellow stone favoured by Georgian architects. What made it different was that it could be made in moulds and, once the mould had been produced, hundreds of exquisitely detailed copies of architectural elements could be made.

Now, Mrs Coade had two great advantages in the artificial stone business. Firstly she was an excellent businesswoman and a fine draughtswoman who could spot a good mould maker a mile off. So when John Bacon hove into view she quickly dispatched Daniel Pincot. Bacon's modelling was so good that soon the finest architects of the era were beating a path to Eleanor's factory in Lambeth where her moulded details were (relatively) cheap, very precise and could be easily arranged into larger elements.

But what made them so attractive to Robert Adam and his like was Eleanor's second advantage. Most artificial stone didn't have a very good reputation. The pieces made by George Davy used on Adam's Brentford gateway to Syon House had crumbled with the first frost, much to Adam's embarrassment. Mrs Coade's stone was different. To a base of ball clay from Dorset she added fine quartz, crushed soda-lime glass, crushed flint and grog – a powder she made from previously fired pieces (usually wasters). The whole lot was mixed up and fired at 1,100 degrees Celsius for four days – a prodigious achievement in an era long before thermo-statically controlled kilns. This made a material so strong and resistant to weathering that even today the surviving pieces look like they're fresh from the mould, despite in some cases having spent 240 years in London's corrosive atmosphere.

Perhaps not surprisingly, Eleanor's business took off. A showroom was opened at the east end of Westminster Bridge and commissions followed from George III, for whom she made the Gothic screen at St George's Chapel, his son, later George IV, and the Royal Naval Hospital in Greenwich, for whom she made the Nelson pediment. The infamous Captain Bligh's tomb was her work, as was the front of Twinings' first tea shop on the Strand. Orders also came from abroad and her work was exported to America, Canada, Brazil, Poland, Russian and South Africa.

After Mrs Coade's death in 1821, her manager, William Croggon, bought the company. So famous had she become by then that she even warranted an obituary in the *Gentleman's Magazine* – an almost unheard of compliment for a businesswoman of the era. Croggon, dogged by the failure of the Duke of York to pay his debts, went bankrupt in 1833. And so the work of Eleanor Coade was largely forgotten. But her legacy quietly lived on.

Next to her old factory stood William Woodington's Lion Brewery, proudly adorned by the two lions it had commissioned from its neighbours. Covered in thick red paint, they survived the Blitz despite the devastation of the surrounding area, and stood guard there until 1950 when the whole area was cleared to make way for the Festival of Britain. Thanks to a personal request from George VI, the lions were removed and cleaned up, one being sent to the All-England Rugby Football Club at Twickenham. The other stood for 15 years outside Waterloo station before wandering back closer to the place of its birth. It now stands on the approaches to Westminster Bridge, not far from Eleanor's showroom, cleaned of its paint and not looking a day older than when it was made.

Episode Forty-Three – The Empty City

If there's one thing that's true of nearly all twentieth-century cities, it's that they're full of people. Often far more people than were envisaged when they were laid out. But there is one which, despite being in the heartlands of Britain, now has absolutely none.

The city of Burlington covers a relatively modest area of 34 acres, served by 60 miles of road. It has four power stations, a reservoir, offices, laundries and a hospital. What makes this city remarkable is that none of this shows up on any map nor is there now a single inhabitant.

Burlington, also known at various times by the codenames Chanticleer, Stockwell, Turnstile and the splendidly sinister '3-site', began as an idea in the early 1950s. Military planners who had been having fun modelling the effects of an all-out nuclear war had estimated that 132 nuclear bombs falling on major British cities would cause hundreds of thousands of casualties and disable all forms of government. In 1955 when William Strath of the Central War Plans Secretariat updated this report to include hydrogen bombs, his estimate for immediate deaths had risen to 12 million (with a further four million serious injuries) – around a third of the population. Plans were considered for providing public shelters but the costs were prohibitive so the answer the Secretariat came up with was to leave the people to their own defences but hide away a core of government officials and ministers in a new, underground city – Burlington. There they could organise the massive retaliation that the doctrine of 'Mutually Assured Destruction' required and then re-establish contact with the outside world and start explaining to the survivors what it had all been about when the radioactive dust settled.

The site chosen was the subterranean 'Spring Quarry', a source of fine Bath stone, near Corsham in Wiltshire and, in 1957, work began in earnest. Burlington was designed on a grid plan as a small-scale mirror of Whitehall, its streets lined with pared down versions of peacetime ministries, complete with ministers and civil servants. This miniature government was provided with food, water and fuel to last three months, after which it was assumed it might be safe to venture outside.

The vast gloomy city was divided into 24 areas where 4,000–6,000 people might live and work whilst the world above was crumbling. Area 21 held the communications offices of the intelligence services that would send out orders to 12 regional bunkers which in turn would be connected to a network of 1,563 monitoring posts manned by the Royal Observer Corps. They would, in turn, relay details of the unfolding devastation back to Burlington.

The heart of the city was Area 17 where politicians and senior officials would live and work around the 'Map Room' in which the course of the war and its aftermath would be plotted. Here the prime minister had the only en-suite bedroom in the entire complex, not that such luxury would probably have interested them. After all, no provision had been made for their, or any other officials', family members, so it was to be a lonely life.

To communicate with those unfortunate enough not to have a place in the bunker, Area 16 contained the BBC broadcasting studio from where the prime minister would update his or her remaining people on the progress of the war whilst Area 8 contained the second largest telephone exchange in Britain which would be manned by GPO staff. Area 12 housed the industrial ovens of the kitchens, making meals from the foodstuffs stockpiled in Area 9 which also housed a fully operational hospital and dental surgery. It was a bleak and utilitarian world at best and rumours that there was a pub – the *Rose and Crown* – were just that – rumours.

It was only with the declassification of the site in 2002 that many civil servants even knew they had a desk reserved for them in Burlington. Had the warning come, they would have been required to immediately leave their tearful families to their fate and set off for their new underground home, alone. Once fully manned, or once no time remained, the blast doors of the site would have been sealed, leaving the inhabitants, protected behind 100-foot-thick concrete walls, to sit out the estimated two-day 'destructive phase'. After this there was an estimated one-month 'survival phase' followed by a 'reconstruction phase' that was rather optimistically scheduled to last just a year. That is assuming there was much left to reconstruct.

Burlington was finally completed in 1961, a year before the Cuban Missile Crisis, and stocked in preparation for the unthinkable event. Fortunately that event never came and when Margaret Thatcher was presented with an estimated bill of £40 million for renovating the site in 1989 she deemed it unnecessary as the threat from the Soviet Bloc had disappeared. The invention of bunker-busting bombs had also made hiding underground a shade pointless. When journalists were finally allowed into the decommissioned structure they found a time capsule – piles of chairs and tables still in their wrappers, thousands of toilet rolls, reams of paper and rows of 1960s telephones still in their boxes. In 1992 the last few maintenance officers left and Burlington, the city that never was, never would be again.

Episode Forty-Four – Shakey Toun

It all began in July of 1597, or rather it probably actually all began millions of years before that, but up to that point no one had seen any good reason to make a note of it. But that year, James Melville decided to take the plunge when he wrote in his diary:

... ther was an erthquak quhilk maid all the north parts of Scotland to trimble from St Jhonftoun throw Athall, Bredalban and all thefe hie lands to Ros, and therin and Kinteall ...

So began a small group of Scottish diarists' role at the heart of the development of a whole new science – seismology. Scotland is perhaps not at the top of your list of 'places where there is a lot of trouble from earthquakes' but the Highland Boundary Fault does from time to time really make the earth there move, particularly around the village of Comrie, in Perthshire. Amongst the first to record this regularly was the Reverend Ralph Taylor of Ochtertyre, just to the east of Comrie, who noticed in the autumn of 1788 that a series of shocks was affecting the area, a fact he considerately communicated to the Royal Society the following year. Taylor, the first of the village's 'earthquake secretaries', kept recording tremors right up until his departure from the area in

1791 when another minister, the Reverend Gilfillan, took over the task of recording every 'trimble' in the area.

And there was a lot of trimbling being done. In his 30 years in the community he recorded at least 68 shocks, noting the effects, the sounds the quakes made and occasionally using this eminent proof of the power of God in his religious writings as well. What Gilfillan did not do however was try to quantify earthquakes in any way which might make them comparable. That wouldn't happen until a sizeable jolt in October 1839, after a long period of dormancy, shook two of the most unlikely heroes in seismology into action.

Peter Macfarlane was more of a postmaster than a scientist, indeed he was Comrie's postmaster, but he set out to analyse and quantify the phenomenon, carrying on the good reverends' work recording each event. But almost immediately he realised that he was lacking a way of quantifying the intensity of each quake. What he came up with was not the world's first seismometer – the laurels for that probably must go to the brilliant Han Dynasty scholar Zhang Heng (AD 78–139). To be fair, Ascanio Filomarino had also got there before Macfarlane in 1795 when the Neapolitan clockmaker had built a device for recording earthquakes which he hoped might give advanced warning of eruptions of Mount Vesuvius. Locals however didn't take to his idea, destroying his seismometer, burning his workshop down and beating him to death for good measure.

Macfarlane was lucky to have less excitable neighbours and was allowed to go about setting up his series of pendulums in relative peace. These simple devices consisted of a weight suspended on a line which would be displaced when the earth shook, as it so frequently did. This, he hoped, would allow him to record both the intensity and direction of the quake. But on what scale was it to record events? To answer this the resourceful postmaster invented one of the very earliest earthquake scales – the Comrie scale – which graded the intensity of quakes on a scale of one to ten, one being 'just sensible' and ten being as powerful as the quake of 23 October 1839, which had first stirred him into action.

To be honest it was perhaps not the most scientific of scales, but Macfarlane didn't have the money or opportunity to travel the world witnessing the effects of other quakes – as Robert Mallet would do in the Great Neapolitan Earthquake of 16 December 1857. But what Macfarlane did do, by his dogged recording, was inspire David Milne-Home, former junior defence lawyer for the notorious murderer William Burke and a man whom an early history of seismology noted was one of '... those who are relieved by ample fortune from the cares of this world'. He, as a man of substance with an interest in geology, persuaded the British Association to appoint a committee to study earthquakes. A flurry of activity followed in which instruments were commissioned and sent to Comrie, all under the care of Macfarlane, to record the extraordinary events regularly occurring in what would become known as 'Shakey Toun'.

However, such interest was not to last. When the quakes began to tail off in 1844 so the great and good turned their back on Comrie. All except Macfarlane of course, who continued through the next 16 unremarkable years to record every minor shock. In 1869 interest flared again as the quakes increased and the committee was reformed. This renewed activity culminated in 1874 in the construction of one of the smallest listed buildings in Scotland – the Earthquake House. This tiny structure, built directly on the bedrock over the Highland Boundary Fault, still houses a replica of its original 'Mallet seismometer'.

Not that Macfarlane would see the device in operation, he died that same year, having become the longest serving and greatest of the Comrie earthquake secretaries, witnessing and recording more tremors than anyone to date – over 300. He has, sadly, not gone down in the annals of seismology as a particularly great name but should you visit the Earthquake House today you will see inside a modern seismograph – the automated descendant of all those earthquake secretaries, a testament to Macfarlane's work and his dutiful successor.

IX

In which we prove that engineers are very much human.

Episode Forty-Five – The Boston Molasses Disaster

Having holidayed in Cornwall I can now vouch for two things. Firstly waves are astonishingly powerful. Secondly, middle-aged historians really shouldn't try to surf. On the upside however the subsequent period of convalescence has given me an excellent opportunity to think about waves in general, and one in particular.

Molasses is the thick treacly by-product of refining sugar and a jolly useful one at that. Not only can you cook with it and eat it, but it can be fermented and distilled to make rum and ethyl alcohol, used to remove rust, added to feedstock and even used in mortar. What it is not generally used for is surfing. However, the terrifying power of a molasses wave has once been tragically experienced.

In the winter of 1918 the United States Industrial Alcohol Company began filling its 15 x 27-metre molasses tank at 529 Commercial Street, Boston, Massachusetts. This vast container, capable of holding 8,700,000 litres of the sticky stuff had been built in 1915, to help sate the USA's ever-increasing appetite for industrial alcohol, largely for use in the munitions business which was undergoing something of a boom thanks to the outbreak of the First World War in Europe. But the owners, in their haste, had made a number of mistakes.

First amongst these was the appointment of Arthur Jell to oversee the construction. Jell was not an engineer, or an architect, but the company reasoned that fabricating a big tank was hardly the same thing as building a railway or a skyscraper, so costly experts were not needed. With no training in engineering, Jell, who was unable to read blueprints, had little idea how to check if the behemoth rising before him would perform as expected and when it was finished he did not even order a simple stress test – filling the tank with water – to check that it would hold up and not leak.

And it did leak; indeed there were reports of it 'weeping' molasses as soon as it was filled. At this point a more risk-averse company might have had it drained and resealed. This one decided to paint the tank brown so no one would notice.

But for all its leakiness, the tank survived for the next four years, creaking and groaning, but providing locals with the occasional bonus of free molasses which they collected from the regular seepage. Then, on 12 January 1919, a molasses tanker from Puerto Rico docked at the wharf and over the next two days finished pumping the tank full to the top. Initially all seemed well, until lunchtime on Wednesday 15th. That day had dawned unusually warm for the time of year. Indeed temperatures had risen over the last two days from a chilly −15.5 degrees Celsius to a relatively balmy 4.4 degrees and the people of this densely populated neighbourhood were out and about their business. Around 12.40 witnesses reported a 'muffled roar' followed by what First World War veterans present said sounded like machine-gun fire. This was actually the rivets shooting out of the rupturing tank. The ground then began shaking 'as though a train were passing' and moments later all hell was let loose.

If a flood of molasses doesn't sound all that terrifying it's worth looking at the statistics. In a matter of seconds nearly nine million litres of the stuff was released onto Commercial Street in a wave which in places reached 4.5 metres high and was reportedly travelling at around 56 km/h. Over 14,000 tonnes of high-speed syrup surged out, exerting a pressure on everything it hit of around 200 kPa. Just the sickly-sweet air blast ahead of the wave threw people off their feet and the wave itself tore the girders off the Boston Elevated Railway and lifted a train from the tracks. People, horses, trucks, carts and whole buildings were lifted onto the wave and tumbled down the street or into the dock.

By this time much of the area was waist deep in black syrup and beneath its glistening surface lay the bodies of 21 people. Another 150 were dragged injured from its clutches. It would take a further 87,000 man-hours to clear the streets and houses and the harbour remained brown with molasses well into the summer. The smell, so local legend has it, has still not entirely gone away.

There was, of course, an inquiry, and fortunately this proved somewhat less of a whitewash than that into the comparable London Beer Flood of 17 October 1814 when 1,470,000 litres of beer from the Meux Brewing Company tore down Tottenham Court Road, drowning eight people in the process. This was declared an 'Act of God' and as such, no one's fault. In the 1919 case the United States Industrial Alcohol Company tried valiantly to claim the Molasses Flood had been caused by anarchists blowing up the tank but in the end blame was placed on the poor construction and insufficient testing of the container. In the end the company paid $600,000 in out-of-court settlements. The damage to property was estimated at what would today be around £100,000,000.

The court case never did get to the bottom of exactly what had caused the tank to fail, however. It was certainly unusually full and some fermentation in the vessel in the rising temperatures over the previous few days may have increased the strain. One scurrilous suggestion is that the company was stocking up on molasses just before the introduction of Prohibition in the hope of making a killing on the rum market. However, as the company only produced industrial alcohol there's no evidence that this is more than just a rumour. In truth the real reason may be no more complex than the company's failure to hire a qualified engineer four years earlier.

Episode Forty-Six – That Sinking Feeling

There's nothing quite like a good piece of mechanical engineering but also there's sadly nothing quite like the instruction manual that usually goes with it. In many cases the fault lies with a particularly excitable (or lazy) translation from a very different language but sometimes it is simply a product of the impenetrable wall that can build up between an engineer and the people who have to use whatever that engineer has made.

If one were looking for a particularly complicated manual then that which accompanied the Kriegsmarine Type VIIC U-boat U-1206 would probably make your list. A magnificent piece of engineering, 568 of this class prowled the world's oceans during the Second World War, bringing terror to the Allied Merchant Marine.

Not that U-1206 had brought much terror herself. Since her launch in December 1943 she hadn't sunk or even damaged a single ship although it was some comfort to her captain, Karl-Adolph Schlitt, that neither had he lost a single crew member. Not that Karl-Adolph had shrunk from battle.

It was now April 1945 – the 14th to be precise – and U-1206 was some 60 metres beneath the North Sea, just ten miles off Peterhead, hunting for British freighters. In the skies above, nervous coastal guard pilots scanned the sea for the telltale sign of a periscope in the water whilst merchantmen hugged the coast hoping to avoid the attentions of these infamous killing machines.

It was at this point that Captain Schlitt made a decision that would change his and his crew's future forever. He decided to go to the lavatory. Now, on the hugely complex U-1206, few things were more complicated than the lavatory. The problems were obviously considerable. Whereas a sailor could go over the side, a soldier could go behind a tree and a pilot could cross his legs and wait until he landed, a submariner, deep beneath the ocean, couldn't really just nip outside to relieve himself. This problem had not daunted the engineers of the Kriegsmarine however and they had come up with a fiendishly complex piece of equipment – the high-pressure toilet. This marvel could be flushed whilst the U-boat was under water but to do so required a complex series of valves to be opened and closed in exactly the right order.

Indeed so complex was the high-pressure toilet that not only did it come with a manual but its own dedicated member of staff. On each boat fitted with the device one of the crew members was given special training in its operation so that they could instruct the crew on how to safely spend a penny, or pfennig in this case, whilst beneath the waves.

What exactly happened in the moments after Captain Schlitt entered the lavatory compartment on that fateful day is a matter of debate. The first version of events, put forward by the captain himself at a reunion many years later in Peterhead, involved faulty equipment; the second, put forward by some other senior members of the crew, involved faulty captaincy.

In the first rendering of events Captain Schlitt finished his business and swiftly operated the complex series of valves which flush the bowl in exactly the correct order. In the second a nervous and slightly embarrassed captain forgot how to operate the mechanism, and, not wishing to look stupid in front of his men, decided against calling out for help from the trained lavatory supervisor, but had a go at remembering as best he could – which wasn't very well.

Either way, and there are those who swear by each version, the result was the same. Levers were pressed, valves were opened but the bowl did not flush. Instead gallons of high-pressure water from the bottom of the North Sea shot up the U-bend and into the compartment, showering the captain in effluent and brine. Nor could he turn the flow off.

Being in a room with an exploding toilet would probably be enough to ruin most people's day in itself, but Karl-Adolph and his crew had the added disadvantages of being under water, in enemy territory and sitting on an unfortunate piece of chemistry. Directly beneath the lavatory was the power bay where the main batteries that powered the vessel when submerged were kept. When the sea-water from the lavatory began flooding this compartment, battery acid and brine started mixing, forming deadly chlorine gas.

There was really no other choice. In an enclosed space, rapidly filling up with poisonous gas, Karl-Adolph gave the order for an emergency surface. The boat shot up and the hatches were opened just in time to see a British coastal patrol plane arc overhead, register their presence and start on a bombing run. As quickly as possible the boat was evacuated and the crew put in rubber life rafts. As the plane aborted its run, perhaps seeing the trouble the vessel was in, Karl-Adolph burnt his orders and gave his last ever command as a U-boat captain – to open the seacocks and scuttle his boat. He, and most of his crew, scrambled ashore and were interned for the rest of the war, the victims of the only submarine ever to be sunk by its own lavatory.

Episode Forty-Seven – All At Sea

There have been enough ludicrous weapons over the years to fill many a book on engineering folly. Perhaps one day I'll tell you about the world's first stealth bomber – the German Linke-Hofman R1, a 1916 leviathan covered in transparent cellulose which glinted so delightfully in the sun that it was perhaps the most visible plane of the whole war. Or would have been if it had ever managed to get fully off the ground. Or I could tell you about those exploding dogs the Russians trained to run under Nazi tanks, but I fear it would incur the wrath of the RSPCA.

I will be staying in Russia however, or at least hovering off its coast. It is undoubtedly the case that some of the very earliest boats were round – coracles being a good example. It is, however, also the case that quite rapidly naval engineers began making boats that were longer than they were wide, with keels, and pointier at the front, to help them to handle the sea. And so ships, particularly fast warships, became progressively pointier and longer right up until 1873.

In that year radical Russian Vice-Admiral Andrei Alexandrovich Popov decided to go back to basics. He already had a formidable reputation as a naval architect, being the man behind the turret-ship *Pyotr Veliky*, one of the most advanced battleships of the age, but now he decided to take a great leap backwards.

Popov's problem was real enough. His ships couldn't carry guns as large as he would like and couldn't get far enough inshore. His answer was to make his new battleships completely circular. After all a ship that was round would have less of an area needing armour plate and hence could be either faster or more heavily protected as well as being more manoeuvrable (not having a keel) and at the same time more stable.

Sadly the Russia Tsar Alexander II also liked the idea so in 1873, work began in the Galerniy dockyards in St Petersburg on the first of two round, flat-bottomed iron-clads. The ships, which were completely circular when viewed from the air, were driven by six engines, each powering one propeller whilst the firepower came from a pair of 11-inch rifled, breech-loading guns. The first such monstrous vessel was the 2,490-ton *Novgorod* and three years later the *Kiev* was laid down at Nicolaiev although her name was soon changed to the *Rear-Admiral Popov* in honour of its far-sighted inventor.

By now you'll be thinking – 'I don't see many circular ships in modern navies. I wonder why?' And you'd be right to wonder. The truth was the idea was an utter disaster. These *'popovkas'*, as the Tsar affectionately called them, had numerous problems. Their bulbous shape meant that even with six engines they were only capable of around 6–7 knots, less than the current on the river Dneiper where they were tested and where they were promptly swept out to sea. Being not only round but flat-bottomed they were also remarkably difficult to keep steady and as they were swept away they spun round repeatedly, making the crew seasick.

Then there were the guns. They were designed to be fired independently, but when one of these monsters was unleashed, the recoil imparted an off-centre force to the ship, making the whole thing spin like a merry-go-round. This made it jolly tricky to aim properly, although it did make them amusing targets in themselves. Twelve bilge keels were added to improve the situation but, frankly, didn't.

When the sailors onboard weren't being spun round like salad, they could boil instead. In the Ukranian summer these shallow, silver dishes which presented a huge surface area to the unforgiving sun, heated up like ovens.

Unbelievably the two *popovkas* did see action in the Russo-Turkish War of 1877–78, although the main action involved the crews being sick over the side as the ships pitched intolerably in even the slightest seas. Retired to serve as permanently moored batteries, they were finally decommissioned in 1903, and then served as store-ships until they were broken up in 1912.

But that wasn't the end of the story. You remember that Tsar Alexander II though the *popovkas* were a jolly good idea. So good in fact that he commissioned a new Royal Yacht, the *Livadia*, based on the same design and built by John Elder & Co. in Clydebank. The Tsar confidently predicted that this novel layout would finally cure the Empress Maria Alexandrovna of her sea-sickness.

Of course the *Livadia* had every modern convenience and was perhaps the most luxurious private vessel of its day. It even boasted a flower garden and a fountain. But it was not a cure for sea-sickness. It took two months to deliver the lumbering, wallowing vessel from Glasgow to southern Spain and the entire crew of seasoned sailors were continuously sick for the whole voyage – even during the subsequent calm passage from Spain to Istanbul.

Fortunately Tsar Alexander II would not have to face his wife's wrath. Just days after the ship was delivered at Sevastopol in 1881, he was assassinated by an anarchist's bomb.

Episode Forty-Eight – Glass Rain

We're all rather familiar with glass skyscrapers these days. Those of you who ever wander around London will have noticed a new one, The Shard, near London Bridge. This startling, mirror-glassed point performs the unusual function of turning a part of Southwark into something more akin to a Flash Gordon cartoon. Personally I'm all in favour of that. But building shiny glass towers did not initially get off to a very good start.

The idea of building vastly tall, glittering minimalist monoliths was something of a goal for the modernist architects such as Mies van der Rohe. His Seagram Building in New York, finished in 1958, was a very good attempt but, due to engineering constraints, it was never the entirely smooth, glass-clad structure he had envisioned. That would have to wait until 1968, when construction began on the John Hancock Tower – officially now called 'Hancock Place'.

Designed by Henry N. Cobb, the tower was to be in the form of a colossal parallelogram, a shape chosen to emphasise the sharp corners of the building, its shorter sides broken up by a vertical notch to further highlight its dizzying height. And at a whisker under 241 metres it was, and remains, the tallest building both in Boston and New England as a whole. The entire thing was then to be clad in mirrored, slightly blue-tinted glass, made from the largest panes available. These, it was hoped, would reflect a lustrous navy blue against the clear city sky.

Construction of such a novel building was, of course, not without its problems. Even during the foundation digging there was trouble with the retaining steel walls of the trench buckling. This rather unfortunately compromised the foundations of the nearby Trinity Church, one of Boston's most treasured monuments. But such things are hardly unusual for major projects and with some shoring up and only some occasional damage to the city's vital utility lines, building went on apace. What is unusual is what happened next. With the tower finished and about to be opened it started raining. Not the normal sort of rain, you understand, but heavy rain. Rain consisting of large numbers of 230 kg panes of shiny mirrored glass which, whenever the wind speed rose to about 70 km/h, would crash down from an extraordinary height onto the sidewalk and buildings below.

This was not really ideal, but what no one was sure of was why it was happening at all. Indeed, so bemused were the engineers that a scale model of the whole of that part of Boston (the Back Bay) was built in a wind tunnel at MIT to try to work out if it was some strange conjunction of prevailing wind and topography that had caused the problem. It had been noticed that in higher winds the entire structure twisted, something that neither architects nor engineers had envisaged when planning the building. The whole building also swayed alarmingly, causing motion-sickness in people at the higher levels. In fact motion sickness was really the least of their problems as a later analysis suggested that under certain wind conditions the whole thing could have fallen down. Thankfully it didn't and the addition of some stiff steel cross bracing (to keep it upright) and a huge tuned mass-damper on the 58th floor (to prevent the motion sickness) brought the building back under control.

None of this however explained the raining glass and so samples were sent across the country to independent laboratories to try to work out if there was anything wrong with the material itself. And there was. The panes were made of three layers – an outer glass layer, a reflective layer and an inner glass layer – bonded together in a sandwich. As the huge glass monolith heated in the daytime sun and then contracted in the cool of night, like some gigantic sundial gnomon, so the air in between the inner and outer layers expanded and contracted – as it does in every piece of double glazing. However, in these massive and very novel panels, the bonding between the inner glass, reflective layer and outer glass had been made so stiff that the entire force of this thermal stressing was not absorbed by the gap but transmitted to the outer pane, which then, rather disconcertingly, pinged off, plummeting onto the street below.

As more panes pinged, so the architects finally resolved that the only answer was to replace every single piece of glass cladding the building – a total of 10,344 panes and at a cost at the time estimated to be around $5–7 million. Whilst the glass was being removed, and to cover the gaps where windows had removed themselves, plywood sheeting was placed over the frames. The modernist dream of a glass monolith was replaced by a joke that Boston was indeed fortunate to have the tallest plywood building in the world. Not exactly what the minimalists had in mind.

Episode Forty-Nine – The Never-Ending Story

I don't know why I didn't think of it before. It could put an end to our dependence on fossil fuels and solve global warming at a stroke. All we need to do is crack perpetual motion. And that's where the idea begins to go rather wrong.

There is a long and generally less than illustrious history of perpetual motion engineering which despite many centuries of failed experiments in the face of the unforgiving laws of thermodynamics shows no sign of abating.

The earliest known attempt to describe a machine that would run forever was made by the twelfth-century Indian mathematician Bhaskara II. His perpetual wheel had containers of mercury or hinged weights around its rim and, as the wheel turned, the weights moved further from or nearer to the centre of the wheel under the influence of gravity. This, supposedly, would ensure that one side of the wheel was always heavier than the other and thus the 'overbalance wheel' (as it became known) would turn of its own accord. Except it didn't – as Bhaskara probably knew.

Sadly not everyone thought Bhaskara's idea was impossible however and the overbalance wheel became a huge hit with mediaeval inventors. The idea even came to the ears of the greatest Renaissance engineer of them all, Leonardo da Vinci, who sketched such devices but, crucially, realised that they were impossible. He reasoned that as each weight moved farther from the rotational axis the gravitational torque was greater but the moment of inertia was simultaneously increased and hence the net gain to the system was zero. In summing up he made one of the more memorable quotes about perpetual motion:

Oh, ye seekers after perpetual motion, how many vain chimeras have you pursued? Go and take your place with the alchemists.

However, few seem to have been put off and the vain chimera hunt was soon back on. Most of these ideas were still variations on the overbalance wheel utilising weights or magnets although Mark Anthony Zimara did invent the 'self-blowing windmill' – a rather handy device by which the turning of the sails powered bellows which produced the wind to turn the sails. Our old friend Cornelius van Drebbel also invented a 'perpetual motion drive' for a clock which was a) not a perpetual motion machine and b) worked. Although it appeared to many at the time to have no energy input, this perpetual winding mechanism was driven by changes in atmospheric pressure and was hence a rather brilliant idea. The concept was eventually perfected by James Cox in the 1760s with the development of Cox's timepiece.

But back to the lunacy. After 1635 there was a flood of other designs as patenting took off, utilising the movement of magnets, water, mercury or in one case cannonballs, through a closed system. Some of their inventors genuinely believed that they could create a machine that would run without an external energy input; many more believed they might be able to make some money from pretending they could create such a machine. John Ernst Worrell Keely managed to liberate $5 million from investors in the late nineteenth century for a machine that, on his death, was found to operate using hidden air-pressure tubes.

The scientific community, keen to distance itself from these shady practices, soon acted and as early as 1775 the Royal Academy in Paris was refusing to deal with proposals concerning perpetual motion as they were, frankly, silly. Not that this stopped people trying. The simple desire to get more out than you put in (what perpetual motion engineers call 'overunity') has meant that these machines are still with us. Not being a lawyer it would be quite wrong of me to comment on the apparent correlation between overunity machines and fraud trials but the idea of creating such a machine has, strangely, had a very positive effect on science. Just hypothesising these impossible engines has helped us to examine the laws of nature. Even the great bongo-playing physicist and Nobel Laureate Richard Feynman imagined such a machine, although in the face of the apparently immutable First Law of Thermodynamics, which all the above machines break, he decided to have a crack at the second.

His hypothetical Brownian Ratchet consisted of a ratcheted cog in a box connected by a massless and frictionless spindle to a paddle in another closed box. The Brownian motion of the gas in the paddle's box ensured that gas molecules were continually and randomly hitting the paddle, but as the ratcheted gear could only turn one way, only those collisions which pushed the paddle in the same direction as the gear, pushed it round. And so the ratchet turned. Except, as Feynman showed, it wouldn't. For the Brownian ratchet to work, the ratchet mechanism itself would have to be so small that Brownian motion would affect it too, knocking the pawl off the gear and hence allowing it to slip back the other way as often as the collisions with the paddle moved it the correct way. The ratchet would not turn but just randomly oscillate, unless there was a temperature gradient between the boxes in which the paddle and ratchet sat.

Even this elegant thought experiment didn't put off some overunity engineers and there is a tragic postscript to Feynman's story. Encouraged by some students, he went to watch a demonstration by one Joseph Papp of an alternative car engine that apparently put out far more energy than was put in. As far as Feynman could see this was a trick, and the 'measuring equipment' which was connected to the mains supply was actually powering the engine. When Papp pulled the plug and the engine continued to run he thought he'd proved his point but Feynman reasoned that there were probably batteries hidden in the 'engine' which had now taken over from the mains. So he asked to hold the plug and continued to do so despite Papp's increasingly urgent requests for its return. Feynman hoped that the batteries would now drain, the engine would stop and he would have uncovered the deception. But, sadly, at this moment the engine exploded, killing one onlooker. What had happened remains something of a mystery to this day. Feynman thought that Papp, who had promised to hand his device over to the Stanford Research Institute for testing, had probably deliberately blown it up to provide an excuse for not sending his dubious engine for testing and for raising further development funding. Papp however sued Feynman for making him lose control of the test by refusing to hand over the plug. Astonishingly, Caltech, Feynman's employer, settled out of court and paid up, proving that there is still something to be made from perpetual motion.

X

Episode Fifty – The Forgotten Engineer

This was my fiftieth column for *E&T* and to celebrate I wanted to give a hooray to one of my engineering heroes who sadly rarely seems to get much recognition.

Working at Bletchley Park during the Second World War was obviously not a good way to become famous. The work was of the utmost secrecy and everyone there signed the Official Secrets Act, so perhaps it's not surprising that so little of what went on there ever leaked out. In recent years there has at least been some recognition given to the genius of Alan Turing, one of the finest mathematical minds of the century, whose work at Bletchley opened the door onto the modern world and who was, in turn, rewarded by his country with persecution that drove him to suicide.

But there was another figure vital to the development of the early computer at Bletchley who, whilst allowed to live out the quiet life Turing never knew, never really received even a hint of recognition for what must go down as one of the most astonishing jobs of practical electrical engineering in history.

Thomas (Tommy) Flowers was not a public schoolboy but the son of a bricklayer with a peculiar love for engineering. This had led him to take an apprenticeship in mechanical engineering at the Royal Arsenal in Woolwich whilst he took night classes in electrical engineering at the University of London. At the age of 21 this in turn led him to the General

Post Office, a booming business just coming to terms with the problem of how to switch between thousands of telephone calls.

It was a problem that intrigued Tommy and in 1930 he moved to the GPO Research Station at Dollis Hill to work on the problem. By the outbreak of the Second World War Flowers was convinced that an entirely electrical system for switching phone calls was possible using valve technology and, in 1942, this idea brought him to the attention of the secret wartime Code and Cypher School at Bletchley Park. The code breakers there had been using electromechanical devices, know as 'bombes', developed by Turing, to help crunch through the huge numbers of combinations needed for cracking German codes. These proved highly effective at attacking the Morse code 'Enigma' messages used by much of the German military, but had hit something of a brick wall when it came to breaking another group of even more important codes.

Since 1942 the German High Command had begun using a new teletype code machine, the Lorenz SZ42. Whilst Enigma machines were well understood, having been available on the open market since 1923, Lorenz SZ42 was a bespoke system offering 5 trillion times more combinations than the 157 trillion already offered by Enigma. The possibility of getting bombes to crunch that many permutations fast enough for the intelligence to have any value, seemed remote.

The man charged with solving this conundrum was Max Newman whose 'Hut F' team was assigned the task of cracking the 'Tunny' intercepts known to emanate from the German High Command. He suggested that high-speed machinery might help crunch the data by comparing messages. And who better to build such machines than the electrical engineers of the Post Office, who were already used to making high-speed switching equipment for automatic exchanges?

So Tommy Flowers was set to work, designing and building these early machines, known as 'Heath Robinsons' after the English cartoonist famous for drawing impossibly complicated devices. But there was a serious problem. Heath Robinsons could not store the contents of the Tunny messages, slowing down processing. Flowers suggested that, provided that it was always left on and undisturbed, he could build a machine in which electronic valves stored the data.

The response from his bosses was not enthusiastic, as it was assumed that the fragile valves would quickly fail if left permanently on. Fortunately this didn't put Flowers off and he managed to devise and develop his system in his spare time and largely at his own expense. His work rate was astonishing – starting in February 1943, his first machine was ready for testing on 8 December of that year. For eight hours the 1,500 thyratron valves performed without failure. The establishment was impressed, but far more importantly Alan Turing was amazed. He realised that this was not simply an electromechanical calculator: its capacity to store data made it something entirely new. Whilst it could not store a program internally (it had to be physically reconfigured for this), nor was it a general-purpose machine, it was undoubtedly the first electronic computer.

The 1-ton, room-sized behemoth was immediately set to work on Tunny intercepts, gaining in the process a new name, thought up by the members of the Women's Royal Naval who operated it – Colossus. Another 2,400-valve machine was immediately ordered and entered service just four days before D-Day, its first work providing crucial decrypts of Tunny messages in the days before the Allied invasion of Europe. Indeed, it was courtesy of Colossus that Eisenhower had a Tunny decrypt in his hand at the moment he gave the order to launch Operation Overlord.

By the end of the war there were ten Colossi working at Bletchley, but with the surrender of Germany their work suddenly became redundant. This presented the government with a problem. These were the most valuable intelligence machines on earth and it was essential that the technology did not get into the hands of potentially hostile powers, but no one was quite sure what to do with them all. Word eventually came down from the government that eight of the machines were to be destroyed and the other two secretly moved to the peacetime government intelligence centre, GCHQ, where one remained in operation until 1960. More valuable than the machines however was Flowers' knowledge of how to build them. Sadly the government didn't seem to see it that way. Keen just to keep him quiet he was given £1,000 compensation, much less than he had personally spent on developing Colossus, and sent back to the Post Office with a reminder that he had signed the Official Secrets Act and hence could never talk about what he had done.

Flowers kept his silence until 1983 and even then, when he was allowed very little recognition for his work, he was banned from explaining how they were employed at Bletchley. As a result, the designer and builder of what was in effect the world's very first electronic computer never received even a fraction of the credit he deserved. Nor of course could he share in the staggering financial returns that commercial computing brought to those who followed in his footsteps unhindered by the Official Secrets Act. Tommy Flowers died in 1998 at the age of 92, largely unrecognised to the last.

Index

111

116